Ben Stacy Jerrik (Ed.)

Synechocystis

Ben Stacy Jerrik (Ed.)

Synechocystis

Cyanobacteria, Heterotroph, Photosynthesis

Part Press

Publisher:
Part Press is a trademark of
International Book Market Service Ltd., 17 Rue Meldrum, Beau Bassin, 1713-01 Mauritius
Email: info@bookmarketservice.com
Website: www.bookmarketservice.com

Published in 2013

Printed in: U.S.A., U.K., Germany. This book was not produced in Mauritius.

ISBN: 978-620-1-54446-8

Contents

Articles

References

Synechocystis

Synechocystis sp. PCC6803	
Scientific classification	
Kingdom:	Bacteria
Phylum:	Cyanobacteria
Order:	Chroococcales
Genus:	*Synechocystis*
Species:	***S.* sp. PCC6803**
Binomial name	
***Synechocystis* sp.**	

Synechocystis is a genus of cyanobacteria primarily represented by the strain *Synechocystis* sp. PCC6803. *S.* sp. PCC6803 lives in freshwater and is capable of both phototrophic growth by oxygenic photosynthesis in sunlight and heterotrophic growth by glycolysis and oxidative phosphorylation during dark periods. It is able to effectively anticipate transitions of light and dark phases by using a circadian clock.

Growth and use as a model organism

Cyanobacteria are model microorganisms for the study of photosynthesis, carbon and nitrogen assimilation, evolution of plant plastids, and adaptability to environmental stresses. *Synechocystis* sp. PCC6803 is one of the most highly studied types of cyanobacteria as it can grow both autotrophically or heterotrophically in the absence of light. It was isolated from a freshwater lake in 1968 and is easily transformed by exogenous DNA. The photosynthetic apparatus is very similar to the one found in plants. This organism also exhibits phototactic movement.

Genome

The genome of *Synechocystis* sp. PCC6803 is contained within one chromosome (3.57 megabases), three small plasmids (5.2, 2.4, and 2.3 kilobases) and four large plasmids (120, 106, 103, and 44 kilobases).[1]

Light-activated heterotrophy

Synechocystis sp. PCC6803 can live completely heterotrophically in the dark, but for yet unknown reasons requires a minimum of 5 to 15 minutes (blue) light per day. This regulatory role of light is intact in both PSI and PSII deficient strains.[2]

NDH-2 is a regulatory quinone:NAD(P)H oxidoreductase[3] global analysis of circadian gene expression indicates that translation genes are expressed at the early subjective day.[4]

Some glycolytic genes are regulated by *sll1330* under light and glucose-supplemented conditions. One of the most important glycolytic genes is fructose-1,6-bisphosphate aldolase (*fbaA*). The mRNA level of *fbaA* is increased under light and glucose-supplemented conditions. But in *Δsll1330*, *fbaA* is not increased under same conditions.[5]

References

Databases

- SynechoNET (http://bioportal.kobic.re.kr/SynechoNET/): integrated protein-protein interaction database of a model cyanobacterium *Synechocystis* sp. PCC 6803. SynechoNET is a specialized cyanobacterial protein-protein interaction database. It shows feasible cyanobacterial domain-domain interactions, as well as their protein level interactions using the model cyanobacterium, *Synechocystis* sp. PCC 6803. Additionally, SynechoNET provides transmembrane topology and domain information, as well as the interaction networks in graphical web interfaces.
- CyanoBase (http://genome.kazusa.or.jp/cyanobase/Synechocystis/): Cyanobacteria carry a complete set of genes for oxygenic photosynthesis, which is the most fundamental life process on the earth. This organism is also interesting from an evolutional viewpoint, for it was born in a very ancient age and has survived in various environments. Chloroplast is believed to have evolved from cyanobacterial ancestors which developed an endosymbiontic relationship with a eukaryotic host cell. CyanoBase provides an easy way of accessing the sequences and all-inclusive annotation data on the structures of the cyanobacterial genomes. This database was originally developed by Makoto Hirosawa, Takakazu Kaneko and Satoshi Tabata, and the current version of CyanoBase has been developed and maintained by Yasukazu Nakamura, Takakazu Kaneko, and Satoshi Tabata at Kazusa DNA Research Institute.
- STRING (http://string.embl.de/): STRING is a database of known and predicted protein-protein interactions.The interactions include direct (physical) and indirect (functional) associations; they are derived from four sources: Genomic Context, High-throughpot Experiments, (Conserved) Coexpression, and Previous Knowledge. The database currently contains 1,513,782 proteins in 373 species. Especially, the database provides interactions for *Synechocystis* sp. PCC 6803.
- cTFbase (http://bioinformatics.zj.cn/cTFbase/index.php): cTFbase contains 1288 putative TFs identified from 21 fully sequenced cyanobacterial genomes. Through its user-friendly interactive interface, users can employ various criteria to retrieve all TF sequences and their detailed annotation information, including sequence features, domain architecture and sequence similarity against the linked databases. Furthermore, cTFbase also provides phylogenetic trees of individual TF family, multiple sequence alignments of the DNA-binding domain and ortholog identification from any selected genomes.

Cyanobacteria

Cyanobacteria	
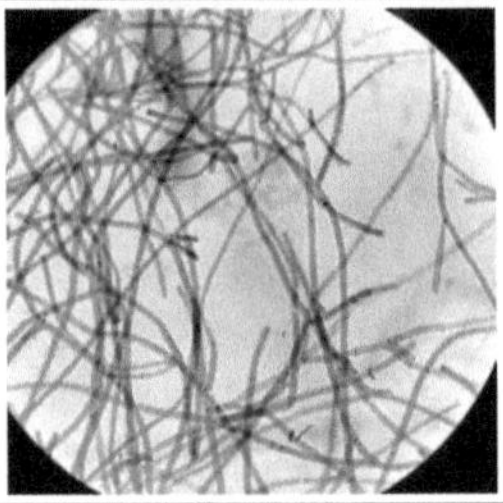	
Tolypothrix sp.	
Scientific classification	
Domain:	Bacteria
Phylum:	**Cyanobacteria**
Orders	
The taxonomy is currently under revision[] [1] • Unicellular forms Chroococcales (suborders-*Chamaesiphonales* and *Pleurocapsales*) • Filamentous (colonial) forms Nostocales (= Hormogonales or Oscillatoriales) • True-branching (budding over multiple axes) Stigonematales	

Cyanobacteria /saɪˌænoʊbækˈtɪəriə/, also known as **blue-green bacteria**, **blue-green algae**, and **Cyanophyta**, is a phylum of bacteria that obtain their energy through photosynthesis.[2] The name "cyanobacteria" comes from the color of the bacteria (Greek: κυανός (kyanós) = blue).

By producing oxygen as a gas as a by-product of photosynthesis, cyanobacteria are thought to have converted the early reducing atmosphere into an oxidizing one, which dramatically changed the composition of life forms on Earth by stimulating biodiversity and leading to the near-extinction of oxygen-intolerant organisms. According to endosymbiotic theory, the chloroplasts found in plants and eukaryotic algae evolved from cyanobacterial ancestors via endosymbiosis.

Ecology

A cyanobacteria bloom near Fiji

Cyanobacteria can be found in almost every terrestrial and aquatic habitat—oceans, fresh water, damp soil, temporarily moistened rocks in deserts, bare rock and soil, and even Antarctic rocks. They can occur as planktonic cells or form phototrophic biofilms. They are found in almost every endolithic ecosystem.[3] A few are endosymbionts in lichens, plants, various protists, or sponges and provide energy for the host. Some live in the fur of sloths, providing a form of camouflage.[4]

Aquatic cyanobacteria are known for their extensive and highly visible blooms that can form in both freshwater and marine environments. The blooms can have the appearance of blue-green paint or scum. These blooms can be toxic, and frequently lead to the closure of recreational waters when spotted. Marine bacteriophages are significant parasites of unicellular marine cyanobacteria.[5]

Characteristics

Nitrogen fixation

Colonies of *Nostoc pruniforme*

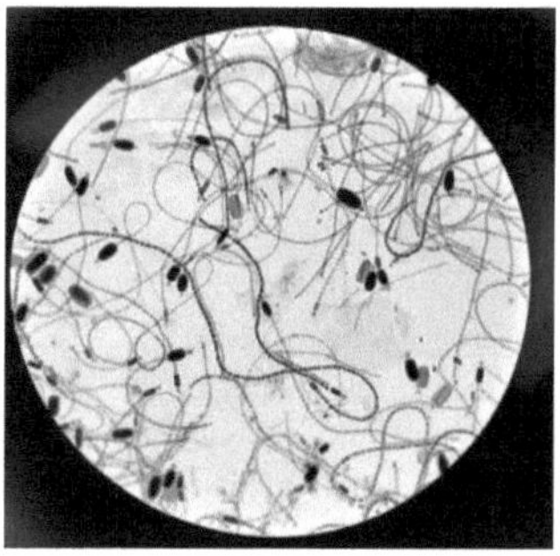

Cylindrospermum sp.

Rice plantations utilize healthy populations of nitrogen-fixing cyanobacteria (*Anabaena*, as symbiotes of the aquatic fern *Azolla*) for use as rice paddy fertilizer.[]

> Cyanobacteria are arguably the most successful group of microorganisms on earth. They are the most genetically diverse; they occupy a broad range of habitats across all latitudes, widespread in freshwater, marine and terrestrial ecosystems, and they are found in the most extreme niches such as hot springs, salt works, and hypersaline bays. Photoautotrophic, oxygen-producing cyanobacteria created the conditions in the planet's early atmosphere that directed the evolution of aerobic metabolism and eukarotic photosynthesis. Cyanobacteria fulfill vital ecological functions in the world's oceans, being important contributors to global carbon and nitrogen budgets.
>
> – Stewart and Falconer[6]

Ecology

Many cyanobacteria form motile filaments of cells, called hormogonia, that travel away from the main biomass to bud and form new colonies elsewhere. The cells in a hormogonium are often thinner than in the vegetative state, and the cells on either end of the motile chain may be tapered. In order to break away from the parent colony, a hormogonium often must tear apart a weaker cell in a filament, called a necridium.

Each individual cell of a cyanobacterium typically has a thick, gelatinous cell wall. They lack flagella, but hormogonia of some species can move about by gliding along surfaces. Many of the multi-cellular filamentous forms of *Oscillatoria* are capable of a waving motion; the filament oscillates back and forth. In water columns some cyanobacteria float by forming gas vesicles, as in archaea. These vesicles are not organelles as such. They are not bounded by lipid membranes but by a protein sheath.

Some of these organisms contribute significantly to global ecology and the oxygen cycle. The tiny marine cyanobacterium *Prochlorococcus* was discovered in 1986 and accounts for more than half of the photosynthesis of the open ocean.[7] Many cyanobacteria even display the circadian rhythms that were once thought to exist only in eukaryotic cells (see bacterial circadian rhythms).

Photosynthesis

Carbon fixation

Cyanobacteria account for 20–30% of Earth's photosynthetic productivity and convert solar energy into biomass-stored chemical energy at the rate of ~450 TW.[] Cyanobacteria utilize the energy of sunlight to drive photosynthesis, a process where the energy of light is used to split water molecules into oxygen, protons, and electrons. While most of the high-energy electrons derived from water are utilized by the cyanobacterial cells for their own needs, a fraction of these electrons are donated to the external environment via electrogenic activity.[] Cyanobacterial electrogenic activity is an important microbiological conduit of solar energy into the biosphere.

Metabolism and organelles

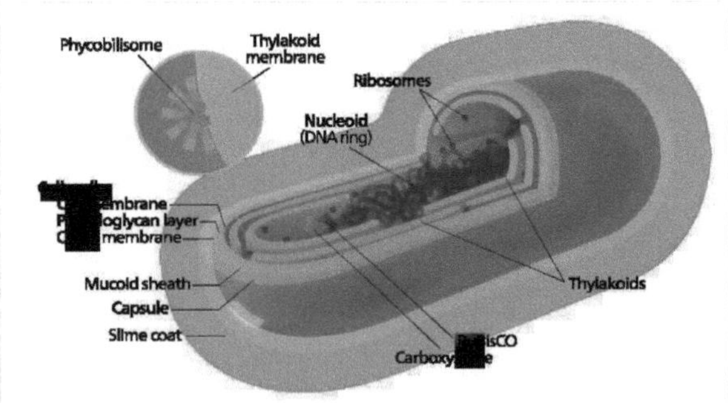

As with any prokaryotic organism, cyanobacteria do not have nuclei or an internal membrane system. However, many species of cyanobacteria have folds on their external membranes which function in photosynthesis. Cyanobacteria get their colour from the bluish pigment phycocyanin, which they use to capture light for photosynthesis. Photosynthesis in cyanobacteria generally uses water as an electron donor and produces oxygen as a by-product, though some may also use hydrogen sulfide a process which occurs among other photosynthetic bacteria such as the purple sulfur bacteria. Carbon dioxide is reduced to form carbohydrates via the Calvin cycle. In most forms the photosynthetic machinery is embedded into folds of the cell membrane, called thylakoids. The large amounts of oxygen in the atmosphere are considered to have been first created by the activities of ancient cyanobacteria. Due to their ability to fix nitrogen in aerobic conditions they are often found as symbionts with a number of other groups of organisms such as fungi (lichens), corals, pteridophytes (*Azolla*), angiosperms (*Gunnera*) etc.

Many cyanobacteria are able to reduce nitrogen and carbon dioxide under aerobic conditions, a fact that may be responsible for their evolutionary and ecological success. The water-oxidizing photosynthesis is accomplished by coupling the activity of photosystem (PS) II and I (Z-scheme). In anaerobic conditions, they are also able to use only PS I—cyclic photophosphorylation—with electron donors other than water (hydrogen sulfide, thiosulphate, or even molecular hydrogen[8]) just like purple photosynthetic bacteria. Furthermore, they share an archaeal property, the ability to reduce elemental sulfur by anaerobic respiration in the dark. Their photosynthetic electron transport shares the same compartment as the components of respiratory electron transport. Their plasma membrane contains only components of the respiratory chain, while the thylakoid membrane hosts an interlinked respiratory and photosynthetic electron transport chain. The terminal oxidases in the thylakoid membrane respiratory/photosynthetic electron transport chain are essential for survival to rapid light changes, although not for dark maintenance under conditions where cells are not light stressed.[9]

Attached to thylakoid membrane, phycobilisomes act as light harvesting antennae for the photosystems. The phycobilisome components (phycobiliproteins) are responsible for the blue-green pigmentation of most cyanobacteria. The variations on this theme are mainly due to carotenoids and phycoerythrins which give the cells the red-brownish coloration. In some cyanobacteria, the color of light influences the composition of phycobilisomes. In green light, the cells accumulate more phycoerythrin, whereas in red light they produce more phycocyanin. Thus the bacteria appear green in red light and red in green light. This process is known as complementary chromatic adaptation and is a way for the cells to maximize the use of available light for photosynthesis.

A few genera, however, lack phycobilisomes and have chlorophyll b instead (*Prochloron*, *Prochlorococcus*, *Prochlorothrix*). These were originally grouped together as the prochlorophytes or chloroxybacteria, but appear to have developed in several different lines of cyanobacteria. For this reason they are now considered as part of the cyanobacterial group.

Relationship to chloroplasts

Gloeobacter

Prochlorococcus

Synechococcus

plastids

all other cyanobacteria

Cladogram showing plastids (chloroplasts and similar) and basal cyanobacteria[10]

Chloroplasts found in eukaryotes (algae and plants) likely evolved from an endosymbiotic relation with cyanobacteria. This endosymbiotic theory is supported by various structural and genetic similarities.[11] Primary chloroplasts are found among the "true plants" or green plants – species ranging from sea lettuce to evergreens and flowers which contain chlorophyll *b* – as well as among the red algae and glaucophytes, marine species which contain phycobilins. It now appears that these chloroplasts probably had a single origin, in an ancestor of the clade called Archaeplastida. Other algae likely took their chloroplasts from these forms by secondary endosymbiosis or ingestion.

Earth history

Stromatolites; (Proterozoic) Zebra River Canyon, Namibia

Stromatolites of fossilized oxygen-producing cyanobacteria have been found from 2.8 billion years ago,[12] possibly from 3.5 billion years ago. The biochemical capacity to use water as the source for electrons in photosynthesis evolved once, in a common ancestor of extant cyanobacteria. The geologic record indicates that this transforming event took place early in our planet's history, at least 2450-2320 million years ago (mya), and probably much earlier. Geobiological interpretation of Archean (>2500 mya) sedimentary rocks remains a challenge; available evidence indicates that life existed 3500 mya, but the question of when oxygenic photosynthesis evolved continues to engender debate and research.

A clear paleontological window on cyanobacterial evolution opened about 2000 mya, revealing an already diverse biota of blue-greens. Cyanobacteria remained principal primary producers throughout the Proterozoic (2500-543 mya), in part because the redox structure of the oceans favored photoautotrophs capable of nitrogen fixation.

The most common cyanobacterial structures in the fossil record are the mound-producing stromatolites and related oncolites. Indeed, these fossil colonies are so common that paleobiology, micropaleontology and paleobotany cite the Pre-Cambrian and Cambrian period as an "age of stromatolites" and an "age of algae."

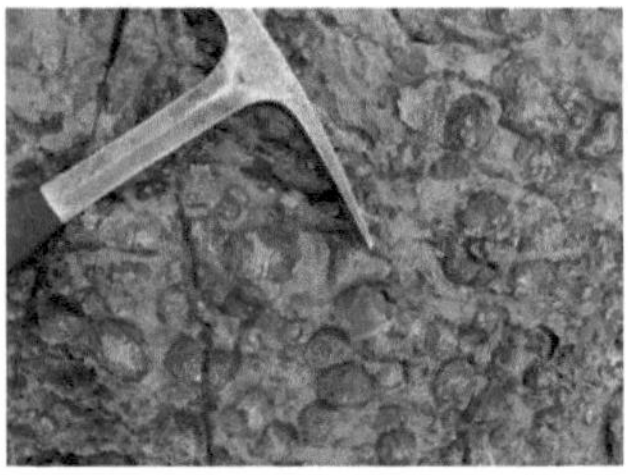

Oncolites; Guilmette Formation (Late Devonian) near Hancock Summit, Pahranagat Range, Nevada

Green algae joined the blue-greens as major primary producers on continental shelves near the end of the Proterozoic, but only with the Mesozoic era (251-65 mya) radiations of dinoflagellates, coccolithophorids, and diatoms did primary production in marine shelf waters take modern form.

Today, the blue-green bacteria remain critical to marine ecosystems as primary producers in oceanic gyres, as agents of biological nitrogen fixation, and—in modified form—as the plastids of marine algae.[]

Classification

Historically, bacteria were first classified as plants constituting the class *Schizomycetes*, which along with the *Schizophyceae* (blue green algae/*Cyanobacteria*) formed the phylum *Schizophyta*.[] then in the phylum *Monera* in the kingdom *Protista* by Haeckel in 1866, comprising *Protogens*, *Protamaeba*, *Vampyrella*, *Protomonae* and *Vibrio*, but not *Nostoc* and other cyanobacteria, which were classified with algae[13] later reclassified as the *Prokaryotes* by Chatton.[14]

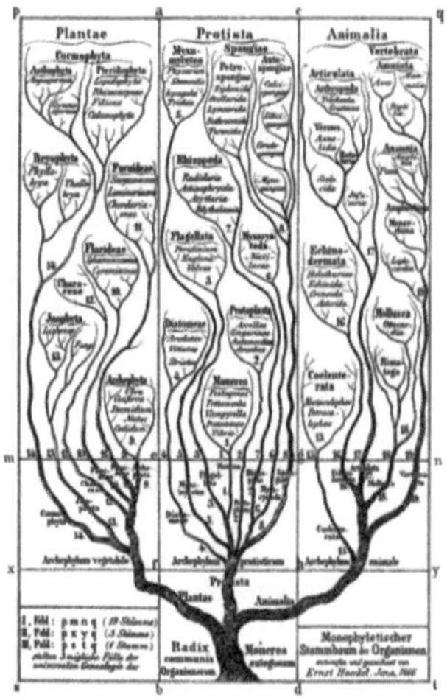

Tree of Life in *Generelle Morphologie der Organismen* (1866). Note the location of the genus *Nostoc* with algae and not with bacteria (kingdom "*Monera*")

The cyanobacteria were traditionally classified by morphology into five sections, referred to by the numerals I-V. The first three – Chroococcales, Pleurocapsales, and Oscillatoriales – are not supported by phylogenetic studies. However, the latter two – Nostocales and Stigonematales – are monophyletic, and make up the heterocystous cyanobacteria. The members of Chroococales are unicellular and usually aggregate in colonies. The classic taxonomic criterion has been the cell morphology and the plane of cell division. In Pleurocapsales, the cells have the ability to form internal spores (baeocytes). The rest of the sections include filamentous species. In Oscillatoriales, the cells are uniseriately arranged and do not form specialized cells (akinetes and heterocysts). In Nostocales and Stigonematales the cells have the ability to develop heterocysts in certain conditions. Stigonematales, unlike Nostocales, includes species with truly branched trichomes. Most taxa included in the phylum or division Cyanobacteria have not yet been validly published under the Bacteriological Code. Except:

- The classes Chroobacteria, Hormogoneae and Gloeobacteria
- The orders Chroococcales, Gloeobacterales, Nostocales, Oscillatoriales, Pleurocapsales and Stigonematales
- The families Prochloraceae and Prochlorotrichaceae
- The genera Halospirulina, Planktothricoides, Prochlorococcus, Prochloron, Prochlorothrix.

Biotechnology and applications

The unicellular cyanobacterium *Synechocystis* sp. PCC6803 was the third prokaryote and first photosynthetic organism whose genome was completely sequenced.[15] It continues to be an important model organism.[16] The smallest genomes have been found in *Prochlorococcus* spp. (1.7 Mb)[17] [18] and the largest in *Nostoc punctiforme* (9 Mb).[19] Those of *Calothrix* spp. are estimated at 12–15 Mb,[20] as large as yeast.

Cyanobacteria cultured in specific media. Cyanobacteria can be helpful in agriculture as they have the capability to fix atmospheric nitrogen to soil.

Some cyanobacteria are sold as food, notably *Aphanizomenon flos-aquae* and *Arthrospira platensis* (Spirulina).[21]

Recent research has suggested the potential application of cyanobacteria to the generation of renewable energy via converting sunlight into electricity. Currently efforts are underway to commercialize algae-based fuels such as diesel, gasoline and jet fuel.[] [22] [23]

Spirulina tablets

Researchers from a company called Algenol have cultured genetically modified cyanobacteria in sea water inside a clear plastic enclosure so that they first make sugar (pyruvate) from CO_2 and the water via photosynthesis. Then, the bacteria secrete ethanol from the cell into the salt water. As the day progresses, and the solar radiation intensifies, ethanol concentrations build up and the ethanol itself evaporates onto the roof of the enclosure. As the sun recedes, evaporated ethanol and water condenses into droplets, which run along the plastic walls and into ethanol collectors, from where it is extracted from the enclosure with the water and ethanol separated outside the enclosure. As of September 2012, Algenol was claiming to have tested its technology on 4 acres in Florida and to have achieved yields of 7,000 US gallons per acre (65,500 litres per hectare) each year. This could potentially meet US demands for ethanol in gasoline in 2025, assuming a B30 blend, from an area of around half the size of California's San Bernardino County, requiring less than one tenth of the area than ethanol from other biomass, such as corn, and only very limited amounts of fresh water.[24]

Cyanobacteria may possess the ability to produce substances that could one day serve as anti-inflammatory agents and combat bacterial infections in humans.[25]

Health risks

Some cyanobacteria produce toxins, called cyanotoxins. These include anatoxin-a, anatoxin-as, aplysiatoxin, cylindrospermopsin, domoic acid, microcystin LR, nodularin R (from *Nodularia*), neosaxitoxin and saxitoxin. Cyanobacteria reproduce explosively under certain conditions. This results in algal blooms, which can become harmful to other species if the cyanobacteria involved produce toxins.

These toxins can be neurotoxins, hepatotoxins, cytotoxins, and endotoxins, and can be toxic and dangerous to humans as well as other animals and marine life in general. Several cases of human poisoning have been documented but a lack of knowledge prevents an accurate assessment of the risks.[26] [27] [28] Recent studies suggest that significant exposure to high levels of some species of cyanobacteria producing toxins such as BMAA can cause amyotrophic lateral sclerosis (ALS, also known as Lou Gehrig's Disease). The Lake Mascoma ALS cluster [29] and Gulf War veteran's cluster are two notable examples.[27] [28] [30]

Dietary supplementation

Microalgae contain substances of high biological value, such as polyunsaturated fatty acids, amino acids (proteins), pigments, antioxidants, vitamins and minerals.[31] Edible blue-green algae reduce the production of pro-inflammatory cytokines by inhibiting NF-κB pathway in macrophages and splenocytes.[32] Consumption of edible blue green algae may also reduce risks of cataracts and age related macular degeneration.[33] It has also shown mitigative effects in animal models of non-alcohol related liver disease, such as steatohepatitis,[34] and Parkinson's disease.[35] Sulfate polysaccharides exhibit immunomodulatory, antitumor, antithrombotic, anticoagulant, anti-mutagenic, anti-inflammatory, antimicrobial, and even antiviral activity against HIV, herpes, and hepatitis.[36] They also improve insulin resistance in HIV.[37] They also protect against aflatoxin and cisplatin chemotherapy induced liver damage.[38] [39] These positive health benefits must be distinguished from non-edible species of algae, which are detrimental to health.[40]

See also

- Anatoxin
- Archean Eon of Earth's prehistory
- Bacterial phyla, the other major lineages of domain Bacteria
- Biofertilizer
- Cyanobiont
- Geological history of oxygen
- The Great Oxygenation Event
- Hypolith
- Microbial mats
- Microalgae
- Phoslock
- Phytoplankton
- Proterozoic Eon of Earth's prehistory
- Sippewissett Microbial Mat
- Stromatolite

References

[5] Photosynthetic viruses keep world's oxygen levels up (http://www.newscientist.com/article/mg20327235.000-photosynthetic-viruses-keep-worlds-oxygen-levels-up.html) *New Scientist* 30 August 2009 by Nora Schultz

[6] Stewart I and Falconer IR (2008) "Cyanobacteria and cyanobacterial toxins" (http://books.google.co.nz/books?id=c6J5hlcjFaAC&pg=PA271&dq="Cyanobacteria+and+cyanobacterial+toxins"&hl=en&ei=rKM4TeKBLIrmvQPKu_SICg&sa=X&oi=book_result&ct=result&resnum=1&ved=0CCkQ6AEwAA#v=onepage&q="Cyanobacteria and cyanobacterial toxins"&f=false) Pages 271–296 in *Oceans and human health: risks and remedies from the seas*, Eds: Walsh PJ, Smith SL and Fleming LE. Academic Press, ISBN 0-12-372584-4.

[7] Steve Nadis, *The Cells That Rule the Seas*, Scientific American, Nov. 2003 (http://sciam.com/article.cfm?chanID=sa022&articleID=0005BE47-0078-1FA8-807883414B7F0000)

[8] Champion Hydrogen-Producing Microbe (http://www.sciencedaily.com/releases/2010/12/101214141932.htm), ScienceDaily, December 15, 2010

[9] Lea-Smith DJ, Ross N, Zori M, Bendall DS, Dennis JS, Scott SA, Smith AG, Howe CJ. Thylakoid terminal oxidases are essential for the cyanobacterium Synechocystis sp. PCC 6803 to survive rapidly changing light intensities. Plant Physiol. 2013 Mar 5. [Epub ahead of print] doi:10.1104/pp.112.210260

[11] Summarised in Cavalier-Smith (2000), *Trends Plant Sci*, **5**: 174–182.

[22] Blue green bacteria may help generate 'green' electricity (http://www.thehindu.com/sci-tech/science/article477049.ece), The Hindu, June 21, 2010

[23] Joule wins key patent for GMO cyanobacteria that create fuels from sunlight, CO2 and water (http://biofuelsdigest.com/bdigest/2010/09/14/joule-wins-key-patent-for-gmo-cyanobacteria-that-create-fuels-from-sunlight-co2-and-water/). Biofuels Digest (2010-09-14). Retrieved on 2011-04-06.

[24] Take it to the Limit: Algenol and rising yields in advanced biofuels (http://www.biofuelsdigest.com/bdigest/2012/09/25/take-it-to-the-limit-algenol-and-rising-yields-in-advanced-biofuels/). Biofuels Digest (2012-09-25). Retrieved on 2012-09-25.

[26] Cyanobacteria, their toxins and health risks (http://www.ncbi.nlm.nih.gov/entrez/query.fcgi?cmd=Retrieve&db=PubMed&list_uids=8830224&dopt=Abstract)

[27] Blue-Green Algae (Cyanobacteria) and their Toxins (http://www.hc-sc.gc.ca/ewh-semt/pubs/water-eau/cyanobacter_e.html)

[28] Harmful Bloom in Lake Atitlán, Guatemala (http://earthobservatory.nasa.gov/NaturalHazards/view.php?id=41487) from NASA Earth Observatory (http://earthobservatory.nasa.gov/), retrieved on 9 January 2010.

[29] Disease Cluster Found at Mascoma Lake , New Hampshire – ALS Research & Treatments – ALS Forum (http://www.als.net/forum/default.aspx?g=posts&t=47305). Als.net. Retrieved on 2011-04-06.

[30] their toxins and health risks (http://quest.mda.org/news/als-research-poison-dirt,)

Further reading

- Gillian Cribbs (1997), *Nature's Superfood: the Blue-Green Algae Revolution*, Newleaf, ISBN 0-7522-0569-2.
- Marshall Savage (1992, 1994), The Millennial Project: Colonizing the Galaxy in Eight Easy Steps, Little, Brown, ISBN 0-316-77163-5.
- Fogg, G.E., Stewart, W.D.P., Fay, P. and Walsby, A.E. (1973), *The Blue-green Algae*, Academic Press, London and New York, ISBN 0-12-261650-2.
- " Architects of the earth's atmosphere (http://www.ucmp.berkeley.edu/bacteria/cyanointro.html)", Introduction to the Cyanobacteria, University of California, Berkeley, 3 Feb. 2006.
- Whitton, B. A., Phylum Cyanophyta (Cyanobacteria), in *The Freshwater Algal Flora of the British Isles*, Cambridge, Cambridge University Press, ISBN 0-521-77051-3.
- .
- Whitton, B. A. and Potts, M. (2000), *The Ecology of Cyanobacteria: their Diversity in Time and Space*, Springer, ISBN 0-7923-4735-8.
- " From Micro-Algae to Blue Oil (http://www.paristechreview.com/2011/12/01/micro-algae-blue-oil/)", ParisTech Review, Dec. 2011.

External links

- What Are Cyanobacteria And What Are Its Types? (http://www.thebigger.com/biology/monera/what-are-cyanobacteria-and-what-are-its-types/)
- Overview of cyanobacteria (http://www.biologie.uni-hamburg.de/b-online/library/webb/BOT311/Cyanobacteria/Cyanobacteria.htm)
- Webserver for Cyanobacteria Research (http://www-cyanosite.bio.purdue.edu)
- CyanoBase (http://genome.kazusa.or.jp/cyanobase)
- Growth Model for the Blue-Green Alga *Anabaena catenula* (http://demonstrations.wolfram.com/GrowthModelForTheBlueGreenAlgaAnabaenaCatenula/) *Wolfram Demonstrations Project*—requires CDF player (free) (http://demonstrations.wolfram.com/download-cdf-player.html)
- Diving an Antarctic Time Capsule Filled With Primordial Life (http://scientistatwork.blogs.nytimes.com/2013/01/31/diving-an-antarctic-time-capsule-filled-with-primordial-life)

Heterotroph

A **heterotroph** (/ˈhɛtərətroʊf/; ἕτερος *heteros* = "another", "different" and τροφή *trophe* = "nutrition") is an organism that cannot fix carbon and uses organic carbon for growth.[1] This contrasts with autotrophs, such as plants and algae, which can use energy from sunlight (photoautotrophs) or inorganic compounds (lithoautotrophs) to produce organic compounds such as carbohydrates, fats, and proteins from inorganic carbon dioxide. These reduced carbon compounds can be used as an energy source by the autotroph and provide the energy in food consumed by heterotrophs.

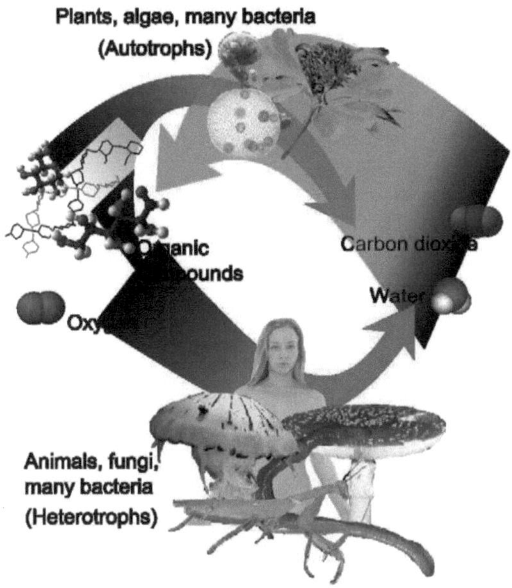

Overview of cycle between autotrophs and heterotrophs

Types

Chemoorganoheterotrophs (or simply organotrophs) exploit reduced carbon compounds as energy sources, like carbohydrates, fats, and proteins from plants and animals. Photoorganoheterotrophs such as Rhodospirillaceae and purple non-sulfur bacteria synthesize organic compounds by utilization of sunlight coupled with oxidation of inorganic substances, including hydrogen sulfide, elemental sulfur, thiosulfate, and molecular hydrogen. They use organic compounds to build structures. They do not fix carbon dioxide and apparently do not have the Calvin cycle.[] Chemolithoheterotrophs can be distinguished from mixotrophs (or facultative chemolithotroph), which can utilize either carbon dioxide or organic carbon as the carbon source.[] []

Heterotrophs, by consuming reduced carbon compounds, are able to use all the energy that they obtain from food for growth and reproduction, unlike autotrophs, which must use some of their energy for carbon fixation. Heterotrophs are unable to make their own food, however, and whether using organic or inorganic energy sources, they can die from a lack of food. This applies not only to animals and fungi but also to bacteria.[]

Ecology

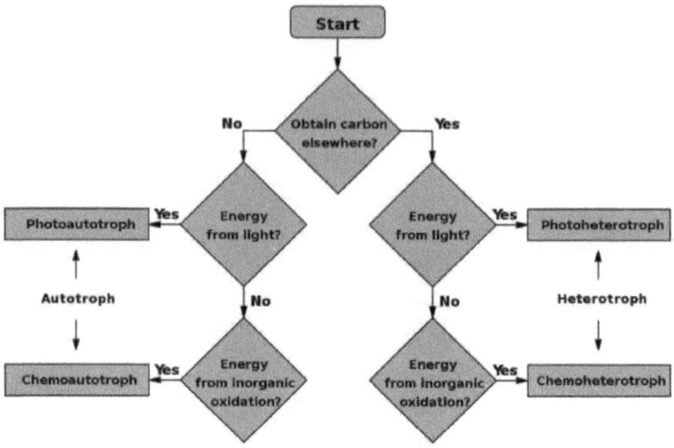

Flowchart to determine if a species is autotroph, heterotroph, or a subtype

Most heterotrophs are chemoorganoheterotrophs (or simply organotrophs) and utilize organic compounds both as a carbon source and an energy source. The term "heterotroph" very often refers to chemoorganoheterotrophs.

Heterotrophs function as consumers in food chains: they obtain organic carbon by eating other heterotrophs or autotrophs. They break down complex organic compounds (e.g., carbohydrates, fats, and proteins) produced by autotrophs into simpler compounds (e.g., carbohydrates into glucose, fats into fatty acids and glycerol, and proteins into amino acids). They release energy by oxidizing carbon and hydrogen atoms present in carbohydrates, lipids, and proteins to carbon dioxide and water, respectively.

Most opisthokonts and prokaryotes are heterotrophic; in particular, all animals and fungi are heterotrophs.[2] Some animals, such as corals, form symbiotic relationships with autotrophs and obtain organic carbon in this way. Furthermore, some parasitic plants have also turned fully or partially heterotrophic, while carnivorous plants consume animals to augment their nitrogen supply while remaining autotrophic.

See also

- Auxotrophy
- Primary nutritional groups
- Saprotrophic nutrition

References

[2] "How Cells Harvest Energy" (http://highered.mcgraw-hill.com/sites/dl/free/0072965819/415836/rav65819_ch07.pdf). McGraw-Hill Higher Education.

Photosynthesis

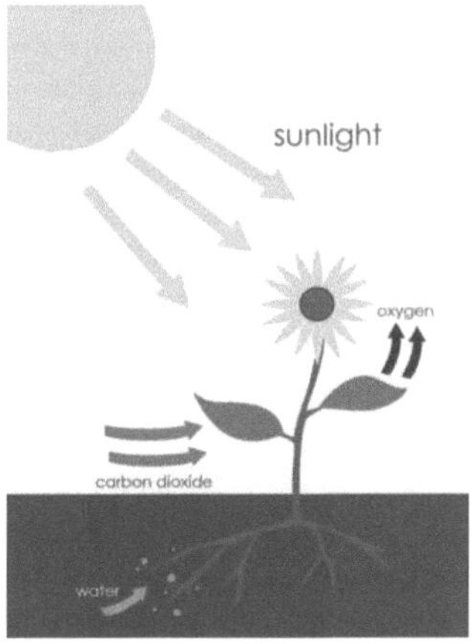

Schematic of photosynthesis in plants. The carbohydrates produced are stored in or used by the plant.

$$6CO_2 + 6H_2O \xrightarrow{Light} C_6H_{12}O_6 + 6O_2$$

Carbon dioxide + Water → Sugar + Oxygen

Overall equation for the type of photosynthesis that occurs in plants

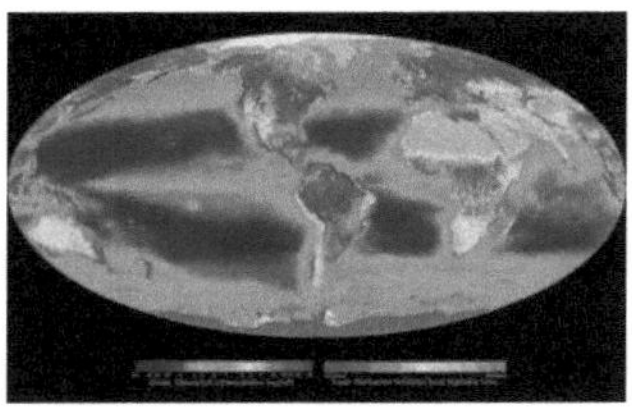

Composite image showing the global distribution of photosynthesis, including both oceanic phytoplankton and terrestrial vegetation. Dark red and blue-green indicate regions of high photosynthetic activity in ocean and land respectively.

Photosynthesis is a process used by plants and other organisms to convert light energy, normally from the sun, into chemical energy that can be used to fuel the organisms' activities. Carbohydrates, such as sugars, are synthesized from carbon dioxide and water (hence the name photosynthesis, from the Greek φῶς, *phōs*, "light", and σύνθεσις, *synthesis*, "putting together"[] [1] [2]). Oxygen is also released, mostly as a waste product. Most plants, most algae, and cyanobacteria perform the process of photosynthesis, and are called photoautotrophs. Photosynthesis maintains atmospheric oxygen levels and supplies all of the organic compounds and most of the energy necessary for all life on Earth.[]

Although photosynthesis is performed differently by different species, the process always begins when energy from light is absorbed by proteins called reaction centres that contain green chlorophyll pigments. In plants, these proteins are held inside organelles called chloroplasts, which are most abundant in leaf cells, while in bacteria they are embedded in the plasma membrane. In these light-dependent reactions, some energy is used to strip electrons from suitable substances such as water, producing oxygen gas. Furthermore, two further compounds are generated: reduced nicotinamide adenine dinucleotide phosphate (NADPH) and adenosine triphosphate (ATP), the "energy currency" of cells.

In plants, algae and cyanobacteria, sugars are produced by a subsequent sequence of light-independent reactions called the Calvin cycle, but some bacteria use different mechanisms, such as the reverse Krebs cycle. In the Calvin cycle, atmospheric carbon dioxide is incorporated into already existing organic carbon compounds, such as ribulose bisphosphate (RuBP).[] Using the ATP and NADPH produced by the light-dependent reactions, the resulting compounds are then reduced and removed to form further carbohydrates such as glucose.

The first photosynthetic organisms probably evolved early in the evolutionary history of life and most likely used reducing agents such as hydrogen or hydrogen sulfide as sources of electrons, rather than water.[] Cyanobacteria appeared later, and the excess oxygen they produced contributed to the oxygen catastrophe,[] which rendered the evolution of complex life possible. Today, the average rate of energy capture by photosynthesis globally is approximately 130 terawatts,[] [] [] which is about six times larger than the current power consumption of human civilization.[] Photosynthetic organisms also convert around 100–115 thousand million metric tonnes of carbon into biomass per year.[] []

Overview

Photosynthetic organisms are photoautotrophs, which means that they are able to synthesize food directly from carbon dioxide and water using energy from light. However, not all organisms that use light as a source of energy carry out photosynthesis, since *photoheterotrophs* use organic compounds, rather than carbon dioxide, as a source of carbon.[] In plants, algae and cyanobacteria, photosynthesis releases oxygen. This is called *oxygenic photosynthesis*. Although there are some differences between oxygenic photosynthesis in plants, algae, and cyanobacteria, the overall process is quite similar in these organisms. However, there are some types of bacteria that carry out anoxygenic photosynthesis, which consumes carbon dioxide but does not release oxygen.

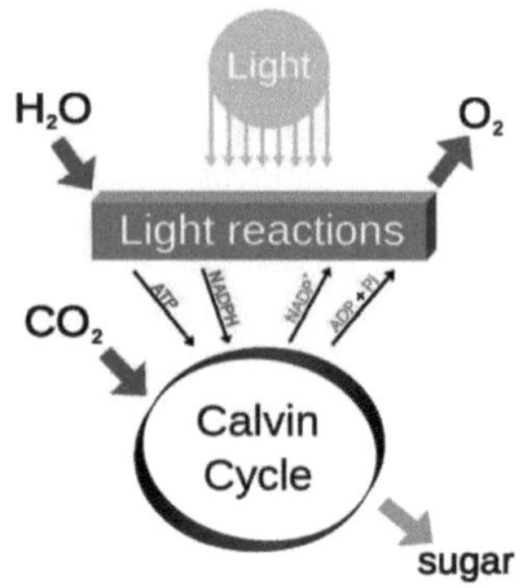

Photosynthesis changes sunlight into chemical energy, splits water to liberate O_2, and fixes CO_2 into sugar.

Carbon dioxide is converted into sugars in a process called carbon fixation. Carbon fixation is an endothermic redox reaction, so photosynthesis needs to supply both a source of energy to drive this process, and the electrons needed to convert carbon dioxide into a carbohydrate. This addition of the electrons is a reduction reaction. In general outline and in effect, photosynthesis is the opposite of cellular respiration, in which glucose and other compounds are oxidized to produce carbon dioxide and water, and to release exothermic chemical energy to drive the organism's metabolism. However, the two processes take place through a different sequence of chemical reactions and in different cellular compartments.

The general equation for photosynthesis is therefore:

$$2n\ CO_2 + 2n\ DH_2 + \text{photons} \rightarrow 2(CH_2O)_n + 2n\ DO$$

Carbon dioxide + electron donor + light energy → carbohydrate + oxidized electron donor

In *oxygenic* photosynthesis water is the electron donor and, since its hydrolysis releases oxygen, the equation for this process is:

$$2n\ CO_2 + 4n\ H_2O + \text{photons} \rightarrow 2(CH_2O)_n + 2n\ O_2 + 2n\ H_2O$$

carbon dioxide + water + light energy → carbohydrate + oxygen + water

Often 2n water molecules are cancelled on both sides, yielding:

$$2n\ CO_2 + 2n\ H_2O + \text{photons} \rightarrow 2(CH_2O)_n + 2n\ O_2$$

carbon dioxide + water + light energy → carbohydrate + oxygen

Other processes substitute other compounds (such as arsenite) for water in the electron-supply role; for example some microbes use sunlight to oxidize arsenite to arsenate:[3] The equation for this reaction is:

$$CO_2 + (AsO_3^{3-}) + \text{photons} \rightarrow (AsO_4^{3-}) + CO$$ []

carbon dioxide + arsenite + light energy → arsenate + carbon monoxide (used to build other compounds in subsequent reactions)

Photosynthesis occurs in two stages. In the first stage, *light-dependent reactions* or *light reactions* capture the energy of light and use it to make the energy-storage molecules ATP and NADPH. During the second stage, the *light-independent reactions* use these products to capture and reduce carbon dioxide.

Most organisms that utilize photosynthesis to produce oxygen use visible light to do so, although at least three use shortwave infrared or, more specifically, far-red radiation.[4]

Photosynthetic membranes and organelles

In photosynthetic bacteria, the proteins that gather light for photosynthesis are embedded within cell membranes, which is the simplest configuration these proteins are arranged.[] However, this membrane may be tightly folded into cylindrical sheets called thylakoids,[] or bunched up into round vesicles called *intracytoplasmic membranes*.[] These structures can fill most of the interior of a cell, giving the membrane a very large surface area and therefore increasing the amount of light that the bacteria can absorb.[]

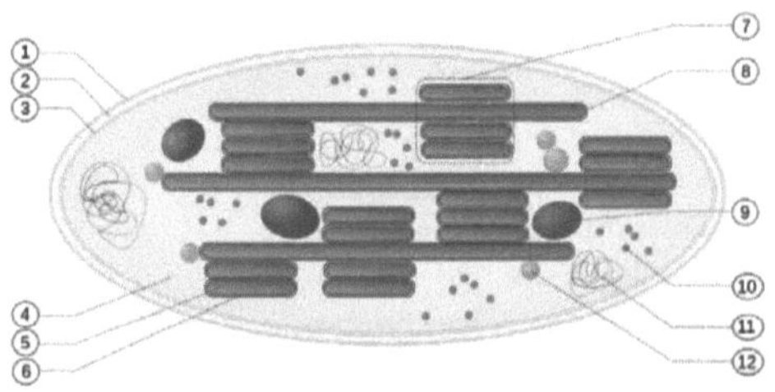

Chloroplast ultrastructure: 1. outer membrane 2. intermembrane space3. inner membrane (1+2+3: envelope) 4. stroma (aqueous fluid) 5. thylakoid lumen (inside of thylakoid) 6. thylakoid membrane 7. granum (stack of thylakoids) 8. thylakoid (lamella) 9. starch 10. ribosome 11. plastidial DNA 12. plastoglobule (drop of lipids)

In plants and algae, photosynthesis takes place in organelles called chloroplasts. A typical plant cell contains about 10 to 100 chloroplasts. The chloroplast is enclosed by a membrane. This membrane is composed of a phospholipid inner membrane, a phospholipid outer membrane, and an intermembrane space between them. Within the membrane is an aqueous fluid called the stroma. The stroma contains stacks (grana) of thylakoids, which are the site of photosynthesis. The thylakoids are flattened disks, bounded by a membrane with a lumen or thylakoid space within it. The site of photosynthesis is the thylakoid membrane, which contains integral and peripheral membrane protein complexes, including the pigments that absorb light energy, which form the photosystems.

Plants absorb light primarily using the pigment chlorophyll, which is the reason that most plants have a green color. Besides chlorophyll, plants also use pigments such as carotenes and xanthophylls.[] Algae also use chlorophyll, but various other pigments are present as phycocyanin, carotenes, and xanthophylls in green algae, phycoerythrin in red algae (rhodophytes) and fucoxanthin in brown algae and diatoms resulting in a wide variety of colors.

These pigments are embedded in plants and algae in special antenna-proteins. In such proteins all the pigments are ordered to work well together. Such a protein is also called a light-harvesting complex.

Although all cells in the green parts of a plant have chloroplasts, most of the energy is captured in the leaves, except in certain species adapted to conditions of strong sunlight and aridity, such as many Euphorbia and Cactus species, whose main photosynthetic organs are their stems. The cells in the interior tissues of a leaf, called the mesophyll, can contain between 450,000 and 800,000 chloroplasts for every square millimeter of leaf. The surface of the leaf is uniformly coated with a water-resistant waxy cuticle that protects the leaf from excessive evaporation of water and decreases the absorption of ultraviolet or blue light to reduce heating. The transparent epidermis layer allows light to pass through to the palisade mesophyll cells where most of the photosynthesis takes place.

Light reactions

In the light reactions, one molecule of the pigment chlorophyll absorbs one photon and loses one electron. This electron is passed to a modified form of chlorophyll called pheophytin, which passes the electron to a quinone molecule, allowing the start of a flow of electrons down an electron transport chain that leads to the ultimate reduction of NADP to NADPH. In addition, this creates a proton gradient across the chloroplast membrane; its dissipation is used by ATP synthase for the concomitant synthesis of ATP.

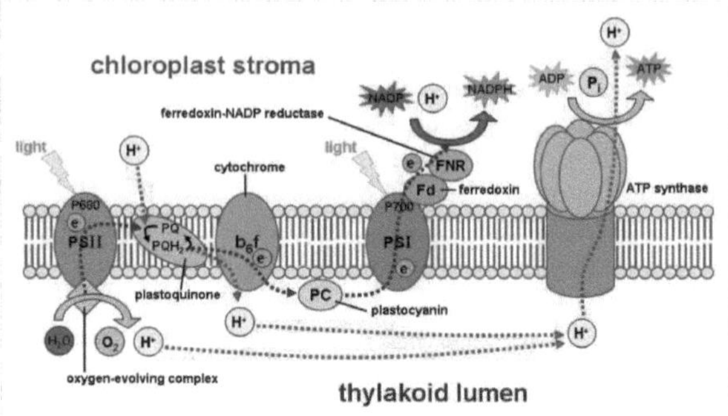

Light-dependent reactions of photosynthesis at the thylakoid membrane

The chlorophyll molecule regains the lost electron from a water molecule through a process called photolysis, which releases a dioxygen (O_2) molecule. The overall equation for the light-dependent reactions under the conditions of non-cyclic electron flow in green plants is:[]

$$2\,H_2O + 2\,NADP^+ + 3\,ADP + 3\,P_i + light \rightarrow 2\,NADPH + 2\,H^+ + 3\,ATP + O_2$$

Not all wavelengths of light can support photosynthesis. The photosynthetic action spectrum depends on the type of accessory pigments present. For example, in green plants, the action spectrum resembles the absorption spectrum for chlorophylls and carotenoids with peaks for violet-blue and red light. In red algae, the action spectrum overlaps with the absorption spectrum of phycobilins for red blue-green light, which allows these algae to grow in deeper waters that filter out the longer wavelengths used by green plants. The non-absorbed part of the light spectrum is what gives photosynthetic organisms their color (e.g., green plants, red algae, purple bacteria) and is the least effective for photosynthesis in the respective organisms.

Z scheme

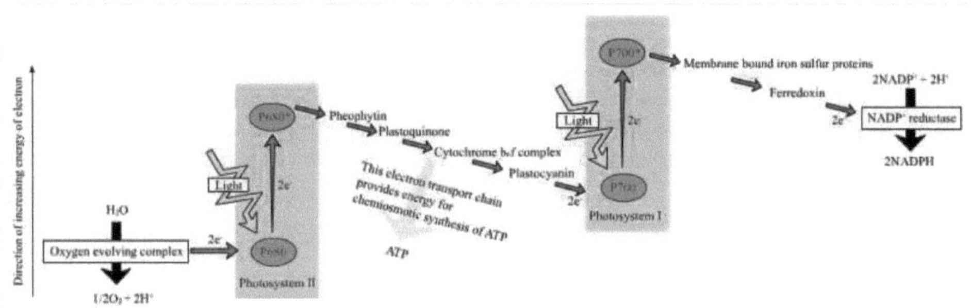

The "Z scheme"

In plants, light-dependent reactions occur in the thylakoid membranes of the chloroplasts and use light energy to synthesize ATP and NADPH. The light-dependent reaction has two forms: cyclic and non-cyclic. In the non-cyclic reaction, the photons are captured in the light-harvesting antenna complexes of photosystem II by chlorophyll and other accessory pigments (see diagram at right). When a chlorophyll molecule at the core of the photosystem II reaction center obtains sufficient excitation energy from the adjacent antenna pigments, an electron is transferred to the primary electron-acceptor molecule, pheophytin, through a process called photoinduced charge separation. These electrons are shuttled through an electron transport chain, the so-called ***Z-scheme*** shown in the diagram, that initially functions to generate a chemiosmotic potential across the membrane. An ATP synthase enzyme uses the chemiosmotic potential to make ATP during photophosphorylation, whereas NADPH is a product of the terminal redox reaction in the *Z-scheme*. The electron enters a chlorophyll molecule in Photosystem I. The electron is excited due to the light absorbed by the photosystem. A second electron carrier accepts the electron, which again is passed down lowering energies of electron acceptors. The energy created by the electron acceptors is used to move hydrogen ions across the thylakoid membrane into the lumen. The electron is used to reduce the co-enzyme NADP, which has functions in the light-independent reaction. The cyclic reaction is similar to that of the non-cyclic, but differs in the form that it generates only ATP, and no reduced NADP (NADPH) is created. The cyclic reaction takes place only at photosystem I. Once the electron is displaced from the photosystem, the electron is passed down the electron acceptor molecules and returns to photosystem I, from where it was emitted, hence the name *cyclic reaction*.

Water photolysis

The NADPH is the main reducing agent in chloroplasts, providing a source of energetic electrons to other reactions. Its production leaves chlorophyll with a deficit of electrons (oxidized), which must be obtained from some other reducing agent. The excited electrons lost from chlorophyll in photosystem I are replaced from the electron transport chain by plastocyanin. However, since photosystem II includes the first steps of the *Z-scheme*, an external source of electrons is required to reduce its oxidized **chlorophyll *a*** molecules. The source of electrons in green-plant and cyanobacterial photosynthesis is water. Two water molecules are oxidized by four successive charge-separation reactions by photosystem II to yield a molecule of diatomic oxygen and four hydrogen ions; the electron yielded in each step is transferred to a redox-active tyrosine residue that then reduces the photoxidized paired-chlorophyll *a* species called P680 that serves as the primary (light-driven) electron donor in the photosystem II reaction center. The oxidation of water is catalyzed in photosystem II by a redox-active structure that contains four manganese ions and a calcium ion; this oxygen-evolving complex binds two water molecules and stores the four oxidizing equivalents that are required to drive the water-oxidizing reaction. Photosystem II is the only known biological enzyme that carries out this oxidation of water. The hydrogen ions contribute to the transmembrane chemiosmotic potential that leads to ATP synthesis. Oxygen is a waste product of light-dependent reactions, but the majority of organisms on Earth use oxygen for cellular respiration, including photosynthetic organisms.[] []

Light-independent reactions

Calvin cycle

In the light-independent (or "dark") reactions, the enzyme RuBisCO captures CO_2 from the atmosphere and in a process that requires the newly formed NADPH, called the Calvin-Benson Cycle, releases three-carbon sugars, which are later combined to form sucrose and starch. The overall equation for the light-independent reactions in green plants is:[] :128

$$3\ CO_2 + 9\ ATP + 6\ NADPH + 6\ H^+ \rightarrow C_3H_6O_3\text{-phosphate} + 9\ ADP + 8\ P_i + 6\ NADP^+ + 3\ H_2O$$

To be more specific, carbon fixation produces an intermediate product, which is then converted to the final carbohydrate products. The carbon skeletons produced by photosynthesis are then variously used to form other organic compounds, such as the building material cellulose, as precursors for lipid and amino acid biosynthesis, or as a fuel in cellular respiration. The latter occurs not only in plants but also in animals when the energy from plants gets passed through a food chain.

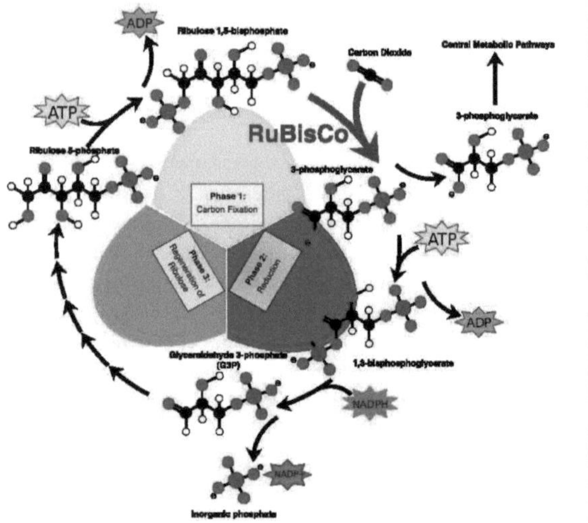

Overview of the Calvin cycle and carbon fixation

The fixation or reduction of carbon dioxide is a process in which carbon dioxide combines with a five-carbon sugar, ribulose 1,5-bisphosphate (RuBP), to yield two molecules of a three-carbon compound, glycerate 3-phosphate (GP), also known as 3-phosphoglycerate (PGA). GP, in the presence of ATP and NADPH from the light-dependent stages, is reduced to glyceraldehyde 3-phosphate (G3P). This product is also referred to as 3-phosphoglyceraldehyde (PGAL) or even as triose phosphate. Triose is a 3-carbon sugar (see carbohydrates). Most (5 out of 6 molecules) of the G3P produced is used to regenerate RuBP so the process can continue (see Calvin-Benson cycle). The 1 out of 6 molecules of the triose phosphates not "recycled" often condense to form hexose phosphates, which ultimately yield sucrose, starch and cellulose. The sugars produced during carbon metabolism yield carbon skeletons that can be used for other metabolic reactions like the production of amino acids and lipids.

Carbon concentrating mechanisms

On land

In hot and dry conditions, plants close their stomata to prevent the loss of water. Under these conditions, CO_2 will decrease, and oxygen gas, produced by the light reactions of photosynthesis, will decrease in the stem, not leaves, causing an increase of photorespiration by the oxygenase activity of ribulose-1,5-bisphosphate carboxylase/oxygenase and decrease in carbon fixation. Some plants have evolved mechanisms to increase the CO_2 concentration in the leaves under these conditions.

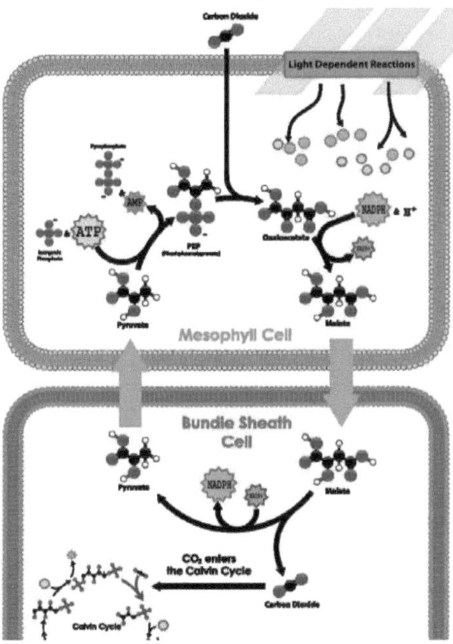

Overview of C4 carbon fixation

C_4 plants chemically fix carbon dioxide in the cells of the mesophyll by adding it to the three-carbon molecule phosphoenolpyruvate (PEP), a reaction catalyzed by an enzyme called PEP carboxylase, creating the four-carbon organic acid oxaloacetic acid. Oxaloacetic acid or malate synthesized by this process is then translocated to specialized bundle sheath cells where the enzyme RuBisCO and other Calvin cycle enzymes are located, and where CO_2 released by decarboxylation of the four-carbon acids is then fixed by RuBisCO activity to the three-carbon sugar 3-phosphoglyceric acids. The physical separation of RuBisCO from the oxygen-generating light reactions reduces photorespiration and increases CO_2 fixation and, thus, photosynthetic capacity of the leaf.[] C_4 plants can produce more sugar than C_3 plants in conditions of high light and temperature. Many important crop plants are C_4 plants, including maize, sorghum, sugarcane, and millet. Plants that do not use PEP-carboxylase in carbon fixation are called C_3 plants because the primary carboxylation reaction, catalyzed by RuBisCO, produces the three-carbon sugar 3-phosphoglyceric acids directly in the Calvin-Benson cycle. Over 90% of plants use C_3 carbon fixation, compared to 3% that use C_4 carbon fixation.[]

Xerophytes, such as cacti and most succulents, also use PEP carboxylase to capture carbon dioxide in a process called Crassulacean acid metabolism (CAM). In contrast to C_4 metabolism, which *physically* separates the CO_2 fixation to PEP from the Calvin cycle, CAM *temporally* separates these two processes. CAM plants have a different leaf anatomy from C_3 plants, and fix the CO_2 at night, when their stomata are open. CAM plants store the CO_2 mostly in the form of malic acid via carboxylation of phosphoenolpyruvate to oxaloacetate, which is then reduced to malate. Decarboxylation of malate during the day releases CO_2 inside the leaves, thus allowing carbon fixation to 3-phosphoglycerate by RuBisCO. Sixteen thousand species of plants use CAM.[]

In water

Cyanobacteria possess carboxysomes, which increase the concentration of CO_2 around RuBisCO to increase the rate of photosynthesis. An enzyme, carbonic anhydrase, located within the carboxysome releases CO_2 from the dissolved hydrocarbonate ions (HCO_3^-). Before the CO_2 diffuses out it is quickly sponged up by RuBisCO, which is concentrated within the carboxysomes. HCO_3^- ions are made from CO_2 outside the cell by another carbonic anhydrase and are actively pumped into the cell by a membrane protein. They cannot cross the membrane as they are charged, and within the cytosol they turn back into CO_2 very slowly without the help of carbonic anhydrase. This causes the HCO_3^- ions to accumulate within the cell from where they diffuse into the carboxysomes.[5] Pyrenoids in algae and hornworts also act to concentrate CO_2 around rubisco.[6]

Order and kinetics

The overall process of photosynthesis takes place in four stages:[]

Stage	Description	Time scale
1	Energy transfer in antenna chlorophyll (thylakoid membranes)	femtosecond to picosecond
2	Transfer of electrons in photochemical reactions (thylakoid membranes)	picosecond to nanosecond
3	Electron transport chain and ATP synthesis (thylakoid membranes)	microsecond to millisecond
4	Carbon fixation and export of stable products	millisecond to second

Efficiency

Plants usually convert light into chemical energy with a photosynthetic efficiency of 3–6%.[7] Actual plants' photosynthetic efficiency varies with the frequency of the light being converted, light intensity, temperature and proportion of carbon dioxide in the atmosphere, and can vary from 0.1% to 8%.[8] By comparison, solar panels convert light into electric energy at an efficiency of approximately 6–20% for mass-produced panels, and above 40% in laboratory devices.

Photosynthesis measurement systems are not designed to directly measure the amount of light absorbed by the leaf. Nevertheless, the light response curves that systems like the LCpro-SD produce, do allow comparisons in photosynthetic efficiency between plants.

Measuring the photosynthetic efficiency of wheat in the field using an LCpro-SD

Evolution

Early photosynthetic systems, such as those from green and purple sulfur and green and purple nonsulfur bacteria, are thought to have been anoxygenic, using various molecules as electron donors. Green and purple sulfur bacteria are thought to have used hydrogen and sulfur as an electron donor. Green nonsulfur bacteria used various amino and other organic acids. Purple nonsulfur bacteria used a variety of nonspecific organic molecules. The use of these molecules is consistent with the geological evidence that the atmosphere was highly reduced at that time.

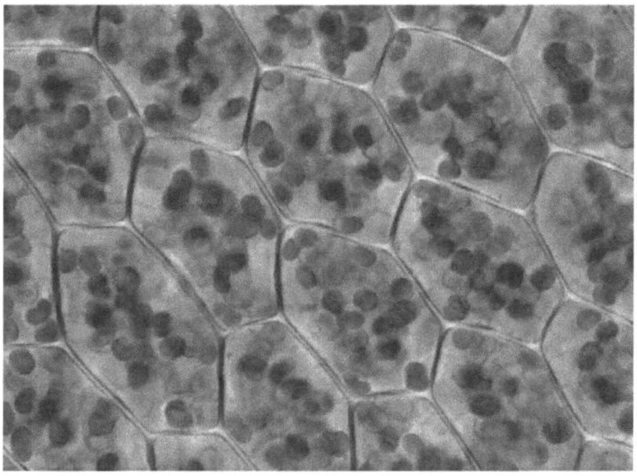

Plant cells with visible chloroplasts (from a moss, *Plagiomnium affine*)

Fossils of what are thought to be filamentous photosynthetic organisms have been dated at 3.4 billion years old.[9] [10]

The main source of oxygen in the atmosphere is oxygenic photosynthesis, and its first appearance is sometimes referred to as the oxygen catastrophe. Geological evidence suggests that oxygenic photosynthesis, such as that in cyanobacteria, became important during the Paleoproterozoic era around 2 billion years ago. Modern photosynthesis in plants and most photosynthetic prokaryotes is oxygenic. Oxygenic photosynthesis uses water as an electron donor, which is oxidized to molecular oxygen (O_2) in the photosynthetic reaction center.

Symbiosis and the origin of chloroplasts

Several groups of animals have formed symbiotic relationships with photosynthetic algae. These are most common in corals, sponges and sea anemones. It is presumed that this is due to the particularly simple body plans and large surface areas of these animals compared to their volumes.[11] In addition, a few marine mollusks *Elysia viridis* and *Elysia chlorotica* also maintain a symbiotic relationship with chloroplasts they capture from the algae in their diet and then store in their bodies. This allows the mollusks to survive solely by photosynthesis for several months at a time.[] [] Some of the genes from the plant cell nucleus have even been transferred to the slugs, so that the chloroplasts can be supplied with proteins that they need to survive.[]

An even closer form of symbiosis may explain the origin of chloroplasts. Chloroplasts have many similarities with photosynthetic bacteria, including a circular chromosome, prokaryotic-type ribosomes, and similar proteins in the photosynthetic reaction center.[] [] The endosymbiotic theory suggests that photosynthetic bacteria were acquired (by endocytosis) by early eukaryotic cells to form the first plant cells. Therefore, chloroplasts may be photosynthetic bacteria that adapted to life inside plant cells. Like mitochondria, chloroplasts still possess their own DNA, separate from the nuclear DNA of their plant host cells and the genes in this chloroplast DNA resemble those in cyanobacteria.[] DNA in chloroplasts codes for redox proteins such as photosynthetic reaction centers. The CoRR Hypothesis proposes that this **Co**-location is required for **R**edox **R**egulation.

Cyanobacteria and the evolution of photosynthesis

The biochemical capacity to use water as the source for electrons in photosynthesis evolved once, in a common ancestor of extant cyanobacteria. The geological record indicates that this transforming event took place early in Earth's history, at least 2450–2320 million years ago (Ma), and, it is speculated, much earlier.[12] Available evidence from geobiological studies of Archean (>2500 Ma) sedimentary rocks indicates that life existed 3500 Ma, but the question of when oxygenic photosynthesis evolved is still unanswered. A clear paleontological window on cyanobacterial evolution opened about 2000 Ma, revealing an already-diverse biota of blue-greens. Cyanobacteria remained principal primary producers throughout the Proterozoic Eon (2500–543 Ma), in part because the redox structure of the oceans favored photoautotrophs capable of nitrogen fixation. Green algae joined blue-greens as major primary producers on continental shelves near the end of the Proterozoic, but only with the Mesozoic (251–65 Ma) radiations of dinoflagellates, coccolithophorids, and diatoms did primary production in marine shelf waters take modern form. Cyanobacteria remain critical to marine ecosystems as primary producers in oceanic gyres, as agents of biological nitrogen fixation, and, in modified form, as the plastids of marine algae.[]

A 2010 study by researchers at Tel Aviv University discovered that the Oriental hornet (*Vespa orientalis*) converts sunlight into electric power using a pigment called xanthopterin. This is the first scientific evidence of a member of the animal kingdom engaging in photosynthesis.[13]

Discovery

Although some of the steps in photosynthesis are still not completely understood, the overall photosynthetic equation has been known since the 19th century.

Jan van Helmont began the research of the process in the mid-17th century when he carefully measured the mass of the soil used by a plant and the mass of the plant as it grew. After noticing that the soil mass changed very little, he hypothesized that the mass of the growing plant must come from the water, the only substance he added to the potted plant. His hypothesis was partially accurate — much of the gained mass also comes from carbon dioxide as well as water. However, this was a signaling point to the idea that the bulk of a plant's biomass comes from the inputs of photosynthesis, not the soil itself.

Joseph Priestley, a chemist and minister, discovered that, when he isolated a volume of air under an inverted jar, and burned a candle in it, the candle would burn out very quickly, much before it ran out of wax. He further discovered that a mouse could similarly "injure" air. He then showed that the air that had been "injured" by the candle and the mouse could be restored by a plant.

In 1778, Jan Ingenhousz, court physician to the Austrian Empress, repeated Priestley's experiments. He discovered that it was the influence of sunlight on the plant that could cause it to revive a mouse in a matter of hours.

In 1796, Jean Senebier, a Swiss pastor, botanist, and naturalist, demonstrated that green plants consume carbon dioxide and release oxygen under the influence of light. Soon afterward, Nicolas-Théodore de Saussure showed that the increase in mass of the plant as it grows could not be due only to uptake of CO_2 but also to the incorporation of water. Thus, the basic reaction by which photosynthesis is used to produce food (such as glucose) was outlined.

Cornelis Van Niel made key discoveries explaining the chemistry of photosynthesis. By studying purple sulfur bacteria and green bacteria he was the first scientist to demonstrate that photosynthesis is a light-dependent redox reaction, in which hydrogen reduces carbon dioxide.

Robert Emerson discovered two light reactions by testing plant productivity using different wavelengths of light. With the red alone, the light reactions were suppressed. When blue and red were combined, the output was much more substantial. Thus, there were two photosystems, one absorbing up to 600 nm wavelengths, the other up to 700 nm. The former is known as PSII, the latter is PSI. PSI contains only chlorophyll a, PSII contains primarily chlorophyll a with most of the available chlorophyll b, among other pigment. These include phycobilins, which are the red and blue pigments of red and blue algae respectively, and fucoxanthol for brown algae and diatoms. The process is most productive when absorption of quanta are equal in both the PSII and PSI, assuring that input energy from the antenna complex is divided between the PSI and PSII system, which in turn powers the photochemistry.[]

Robert Hill thought that a complex of reactions consisting of an intermediate to cytochrome b_6 (now a plastoquinone), another is from cytochrome f to a step in the carbohydrate-generating mechanisms. These are linked by plastoquinone, which does require energy to reduce cytochrome f for it is a sufficient reductant. Further experiments to prove that the oxygen developed during the photosynthesis of green plants came from water, were performed by Hill in 1937 and 1939. He showed that isolated chloroplasts give off oxygen in the presence of unnatural reducing agents like iron oxalate, ferricyanide or benzoquinone after exposure to light. The Hill reaction is as follows:

$$2\ H_2O + 2\ A + (\text{light, chloroplasts}) \rightarrow 2\ AH_2 + O_2$$

where A is the electron acceptor. Therefore, in light, the electron acceptor is reduced and oxygen is evolved.

Samuel Ruben and Martin Kamen used radioactive isotopes to determine that the oxygen liberated in photosynthesis came from the water.

Melvin Calvin works in his photosynthesis laboratory.

Melvin Calvin and Andrew Benson, along with James Bassham, elucidated the path of carbon assimilation (the photosynthetic carbon reduction cycle) in plants. The carbon reduction cycle is known as the Calvin cycle, which ignores the contribution of Bassham and Benson. Many scientists refer to the cycle as the Calvin-Benson Cycle, Benson-Calvin, and some even call it the Calvin-Benson-Bassham (or CBB) Cycle.

Nobel Prize-winning scientist Rudolph A. Marcus was able to discover the function and significance of the electron transport chain.

Otto Heinrich Warburg and Dean Burk discovered the I-quantum photosynthesis reaction that splits the CO_2, activated by the respiration.[14]

Louis N.M. Duysens and Jan Amesz discovered that chlorophyll a will absorb one light, oxidize cytochrome f, chlorophyll a (and other pigments) will absorb another light, but will reduce this same oxidized cytochrome, stating the two light reactions are in series.

Factors

There are three main factors affecting photosynthesis and several corollary factors. The three main are:

- Light irradiance and wavelength
- Carbon dioxide concentration
- Temperature.

The leaf is the primary site of photosynthesis in plants.

Light intensity (irradiance), wavelength and temperature

In the early 20th century, Frederick Blackman and Gabrielle Matthaei investigated the effects of light intensity (irradiance) and temperature on the rate of carbon assimilation.

- At constant temperature, the rate of carbon assimilation varies with irradiance, initially increasing as the irradiance increases. However, at higher irradiance, this relationship no longer holds and the rate of carbon assimilation reaches a plateau.
- At constant irradiance, the rate of carbon assimilation increases as the temperature is increased over a limited range. This effect is seen only at high irradiance levels. At low irradiance, increasing the temperature has little influence on the rate of carbon assimilation.

These two experiments illustrate vital points: First, from research it is known that, in general, photochemical reactions are not affected by temperature. However, these experiments clearly show that temperature affects the rate of carbon assimilation, so there must be two sets of reactions in the full process of carbon assimilation. These are, of course, the light-dependent 'photochemical' stage and the light-independent, temperature-dependent stage. Second, Blackman's experiments illustrate the concept of limiting factors. Another limiting factor is the wavelength of light. Cyanobacteria, which reside several meters underwater, cannot receive the correct wavelengths required to cause photoinduced charge separation in conventional photosynthetic pigments. To combat this problem, a series of proteins with different pigments surround the reaction center. This unit is called a phycobilisome.

Carbon dioxide levels and photorespiration

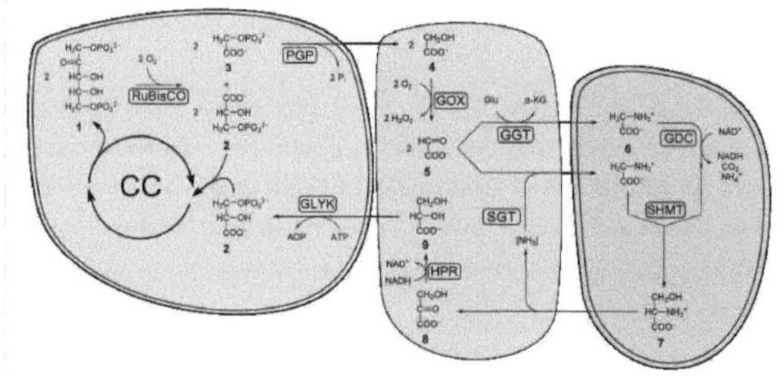

Photorespiration

As carbon dioxide concentrations rise, the rate at which sugars are made by the light-independent reactions increases until limited by other factors. RuBisCO, the enzyme that captures carbon dioxide in the light-independent reactions, has a binding affinity for both carbon dioxide and oxygen. When the concentration of carbon dioxide is high, RuBisCO will fix carbon dioxide. However, if the carbon dioxide concentration is low, RuBisCO will bind oxygen instead of carbon dioxide. This process, called photorespiration, uses energy, but does not produce sugars.

RuBisCO oxygenase activity is disadvantageous to plants for several reasons:

1. One product of oxygenase activity is phosphoglycolate (2 carbon) instead of 3-phosphoglycerate (3 carbon). Phosphoglycolate cannot be metabolized by the Calvin-Benson cycle and represents carbon lost from the cycle. A high oxygenase activity, therefore, drains the sugars that are required to recycle ribulose 5-bisphosphate and for the continuation of the Calvin-Benson cycle.
2. Phosphoglycolate is quickly metabolized to glycolate that is toxic to a plant at a high concentration; it inhibits photosynthesis.
3. Salvaging glycolate is an energetically expensive process that uses the glycolate pathway, and only 75% of the carbon is returned to the Calvin-Benson cycle as 3-phosphoglycerate. The reactions also produce ammonia (NH_3), which is able to diffuse out of the plant, leading to a loss of nitrogen.

A highly simplified summary is:

$$2 \text{ glycolate} + \text{ATP} \rightarrow \text{3-phosphoglycerate} + \text{carbon dioxide} + \text{ADP} + NH_3$$

The salvaging pathway for the products of RuBisCO oxygenase activity is more commonly known as photorespiration, since it is characterized by light-dependent oxygen consumption and the release of carbon dioxide.

History of the concept

In 1893, Charles Reid Barnes proposed two terms, *photosyntax* and *photosynthesis*, for the biological process of *synthesis of complex carbon compounds out of carbonic acid, in the presence of chlorophyll, under the influence of light*. Over time, the term *photosynthesis* came into common usage as the term of choice. Later discovery of anoxygenic photosynthetic bacteria and photophosphorylation necessitated redefinition of the term.[15]

See also

- Jan Anderson (scientist)
- Artificial photosynthesis
- Calvin-Benson cycle
- Carbon fixation
- Cellular respiration
- Chemosynthesis
- Light-dependent reaction
- Photobiology
- Photoinhibition
- Photosynthetic reaction center
- Photosynthetically active radiation
- Photosystem
- Photosystem I
- Photosystem II
- Quantum biology
- Red edge
- Vitamin D

References

[1] φῶς (http://www.perseus.tufts.edu/hopper/text?doc=Perseus:text:1999.04.0057:entry=fw=s2). Liddell, Henry George; Scott, Robert; *A Greek–English Lexicon* at the Perseus Project

[2] σύνθεσις (http://www.perseus.tufts.edu/hopper/text?doc=Perseus:text:1999.04.0057:entry=su/nqesis). Liddell, Henry George; Scott, Robert; *A Greek–English Lexicon* at the Perseus Project

[3] *Anaerobic Photosynthesis*, Chemical & Engineering News, **86**, 33, August 18, 2008, p. 36

[7] Chapter 1 – Biological energy production">

[8] Govindjee, What is photosynthesis? (http://www.life.uiuc.edu/govindjee/whatisit.htm)

[9] Photosynthesis got a really early start (http://www.newscientist.com/article/mg18424671.600-photosynthesis-got-a-really-early-start.html), New Scientist, 2 October 2004

[10] Revealing the dawn of photosynthesis (http://www.newscientist.com/article/mg19125654.200-revealing-the-dawn-of-photosynthesis.html), New Scientist, 19 August 2006

[13] Error: Bad DOI specified: 10.1007/s00114-010-0728-1

[14] Otto Warburg – Biography (http://nobelprize.org/nobel_prizes/medicine/laureates/1931/warburg.html). Nobelprize.org (1970-08-01). Retrieved on 2011-11-03.

[15] Gest, Howard (2002). *History of the word photosynthesis and evolution of its definition.* (http://link.springer.com/article/10.1023/A:1020419417954#) Photosynthesis Research 73(1-3): 7-10.

Further reading

External links

- A collection of photosynthesis pages for all levels from a renowned expert (Govindjee) (http://www.life.uiuc.edu/govindjee/linksPSed.htm)
- In depth, advanced treatment of photosynthesis, also from Govindjee (http://www.life.uiuc.edu/govindjee/paper/gov.html)
- Science Aid: Photosynthesis (http://scienceaid.co.uk/biology/biochemistry/photosynthesis.html) Article appropriate for high school science
- Metabolism, Cellular Respiration and Photosynthesis – The Virtual Library of Biochemistry and Cell Biology (http://www.biochemweb.org/metabolism.shtml)
- Overall examination of Photosynthesis at an intermediate level (http://www.chemsoc.org/networks/learnnet/cfb/Photosynthesis.htm)
- Overall Energetics of Photosynthesis (http://www.life.uiuc.edu/govindjee/photosynBook.html)
- Photosynthesis Discovery Milestones (http://www.juliantrubin.com/bigten/photosynthesisexperiments.html) – experiments and background
- The source of oxygen produced by photosynthesis (http://bcs.whfreeman.com/thelifewire/content/chp08/0802001.html) Interactive animation, a textbook tutorial
- Photosynthesis – Light Dependent & Light Independent Stages (http://www.biology-innovation.co.uk/pages/plant-biology-ecology/photosynthesis/)
- Khan Academy, video introduction (http://www.khanacademy.org/video/photosynthesis?playlist=Biology)

Glycolysis

Glycolysis (from *glycose*, an older term[1] for glucose + *-lysis* degradation) is the metabolic pathway that converts glucose $C_6H_{12}O_6$, into pyruvate, $CH_3COCOO^- + H^+$. The free energy released in this process is used to form the high-energy compounds ATP (adenosine triphosphate) and NADH (reduced nicotinamide adenine dinucleotide).[2]

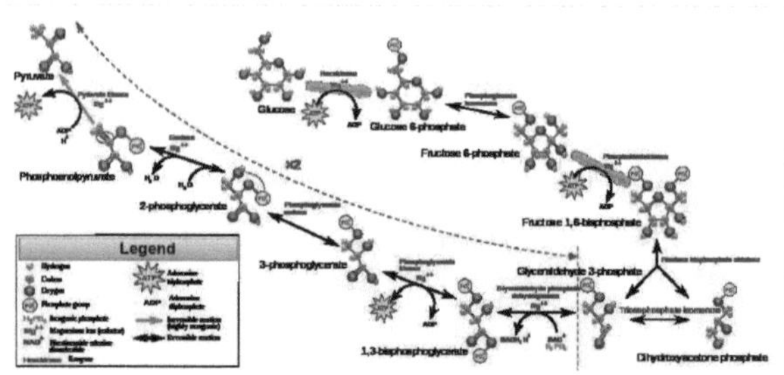

Glycolysis overview

Glycolysis is a purely anaerobic reaction. While it can happen in the presence of oxygen, oxygen is never involved in the reaction, nor does it alter it. Terms like "aerobic glycolysis" are a misnomer, but are sometimes used to describe the environment of the cell and how it affects the metabolic breakdown of the pyruvate product.

Glycolysis is a determined sequence of ten reactions involving ten intermediate compounds (one of the steps involves two intermediates). The intermediates provide entry points to glycolysis. For example, most monosaccharides, such as fructose, glucose, and galactose, can be converted to one of these intermediates. The intermediates may also be directly useful. For example, the intermediate dihydroxyacetone phosphate (DHAP) is a source of the glycerol that combines with fatty acids to form fat.

It occurs, with variations, in nearly all organisms, both aerobic and anaerobic. The wide occurrence of glycolysis indicates that it is one of the most ancient known metabolic pathways.[3] It occurs in the cytosol of the cell.

The most common type of glycolysis is the *Embden–Meyerhof–Parnas (EMP pathway)*, which was first discovered by Gustav Embden, Otto Meyerhof, and Jakub Karol Parnas. Glycolysis also refers to other pathways, such as the *Entner–Doudoroff pathway* and various heterofermentative and homofermentative pathways. However, the discussion here will be limited to the Embden–Meyerhof–Parnas pathway.[4]

The entire glycolysis pathway can be separated into two phases:[5]

1. The Preparatory Phase – in which ATP is consumed and is hence also known as the investment phase
2. The Pay Off Phase – in which ATP is produced.

Overview

The overall reaction of glycolysis is:

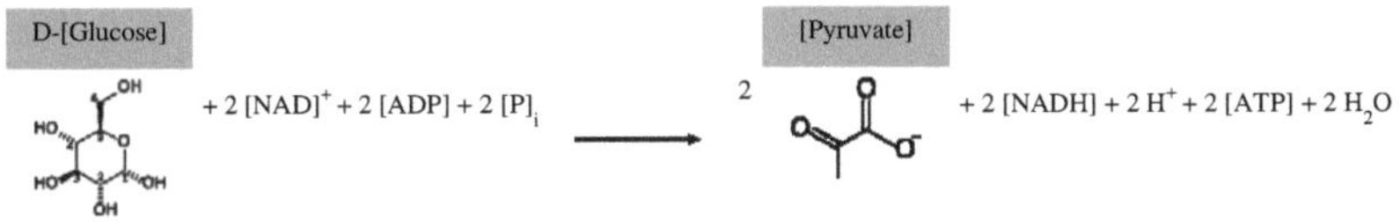

The use of symbols in this equation makes it appear unbalanced with respect to oxygen atoms, hydrogen atoms, and charges. Atom balance is maintained by the two phosphate (P_i) groups:[6]

- each exists in the form of a hydrogen phosphate anion (HPO_4^{2-}), dissociating to contribute 2 H^+ overall
- each liberates an oxygen atom when it binds to an ADP (adenosine diphosphate) molecule, contributing 2 O overall

Charges are balanced by the difference between ADP and ATP. In the cellular environment, all three hydroxy groups of ADP dissociate into $-O^-$ and H^+, giving ADP^{3-}, and this ion tends to exist in an ionic bond with Mg^{2+}, giving $ADPMg^-$. ATP behaves identically except that it has four hydroxy groups, giving $ATPMg^{2-}$. When these differences along with the true charges on the two phosphate groups are considered together, the net charges of –4 on each side are balanced.

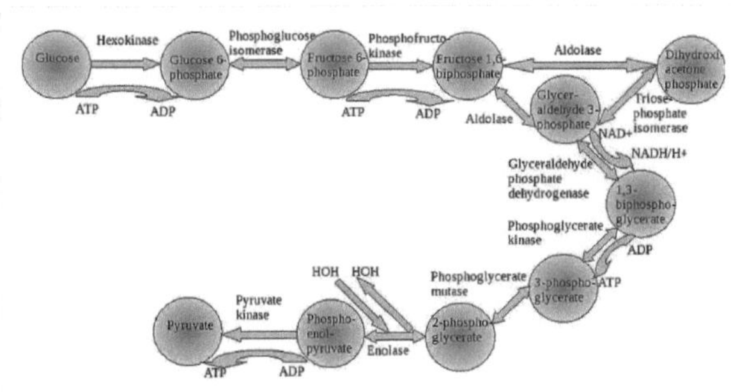

Glycolysis

For simple fermentations, the metabolism of one molecule of glucose to two molecules of pyruvate has a net yield of two molecules of ATP. Most cells will then carry out further reactions to 'repay' the used NAD^+ and produce a final product of ethanol or lactic acid. Many bacteria use inorganic compounds as hydrogen acceptors to regenerate the NAD^+.

Cells performing aerobic respiration synthesize much more ATP, but not as part of glycolysis. These further aerobic reactions use pyruvate and NADH + H^+ from glycolysis. Eukaryotic aerobic respiration produces approximately 34 additional molecules of ATP for each glucose molecule, however most of these are produced by a vastly different mechanism to the substrate-level phosphorylation in glycolysis.

The lower-energy production, per glucose, of anaerobic respiration relative to aerobic respiration, results in greater flux through the pathway under hypoxic (low-oxygen) conditions, unless alternative sources of anaerobically oxidizable substrates, such as fatty acids, are found.

Elucidation of the pathway

In 1860, Louis Pasteur discovered that microorganisms are responsible for fermentation. In 1897, Eduard Buchner found that *extracts* of certain cells can cause fermentation. In 1905, Arthur Harden and William Young along with Nick Sheppard determined that a heat-sensitive high-molecular-weight subcellular fraction (the enzymes) and a heat-insensitive low-molecular-weight cytoplasm fraction (ADP, ATP and NAD^+ and other cofactors) are required together for fermentation to proceed. The details of the pathway were eventually determined by 1940, with a major input from Otto Meyerhof and some years later by Luis Leloir. The biggest difficulties in determining the intricacies of the pathway were due to the very short lifetime and low steady-state concentrations of the intermediates of the fast glycolytic reactions.

Sequence of reactions

Preparatory phase

The first five steps are regarded as the preparatory (or investment) phase, since they consume energy to convert the glucose into two three-carbon sugar phosphates[5] (G3P).

The first step in glycolysis is phosphorylation of glucose by a family of enzymes called hexokinases to form glucose 6-phosphate (G6P). This reaction consumes ATP, but it acts to keep the glucose concentration low, promoting continuous transport of glucose into the cell through the plasma membrane transporters. In addition, it blocks the glucose from leaking out – the cell lacks transporters for G6P, and free diffusion out of the cell is prevented due to the charged nature of G6P. Glucose may alternatively be formed from the phosphorolysis or hydrolysis of intracellular starch or glycogen.

In animals, an isozyme of hexokinase called glucokinase is also used in the liver, which has a much lower affinity for glucose (K_m in the vicinity of normal glycemia), and differs in regulatory properties. The different substrate affinity and alternate regulation of this enzyme are a reflection of the role of the liver in maintaining blood sugar levels.

Cofactors: Mg^{2+}

D-Glucose (**Glc**)	Hexokinase (**HK**) *a transferase*	α-D-Glucose-6-phosphate (**G6P**)

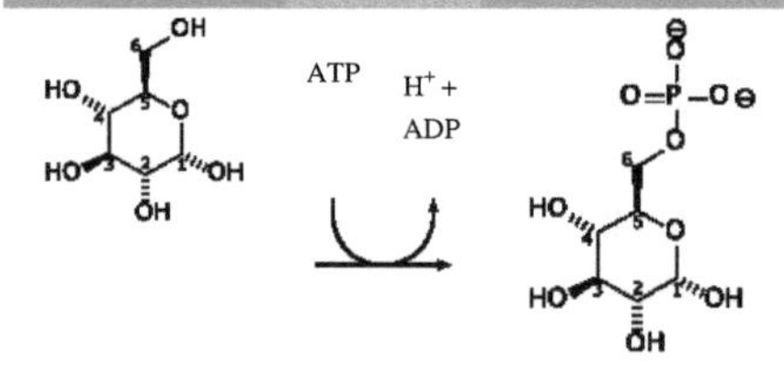

G6P is then rearranged into fructose 6-phosphate (F6P) by glucose phosphate isomerase. Fructose can also enter the glycolytic pathway by phosphorylation at this point.

The change in structure is an isomerization, in which the G6P has been converted to F6P. The reaction requires an enzyme, phosphohexose isomerase, to proceed. This reaction is freely reversible under normal cell conditions. However, it is often driven forward because of a low concentration of F6P, which is constantly consumed during the next step of glycolysis. Under conditions of high F6P concentration, this reaction readily runs in reverse. This phenomenon can be explained through Le Chatelier's Principle. Isomerization to a keto sugar is necessary for carbanion stabilization in the fourth reaction step (below).

α-D-Glucose 6-phosphate (**G6P**)	Phosphoglucose isomerase *an isomerase*	β-D-Fructose 6-phosphate (**F6P**)

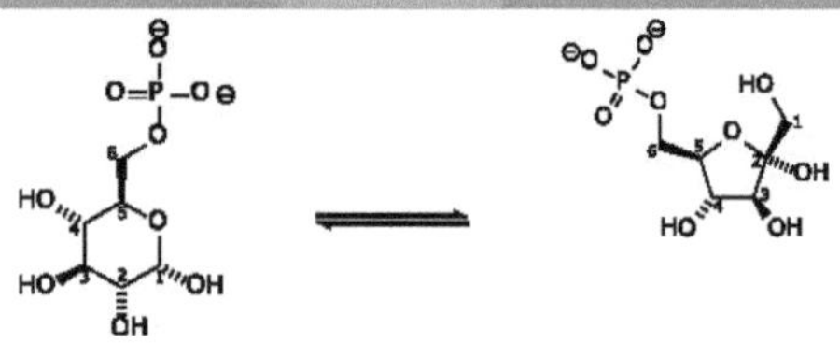

The energy expenditure of another ATP in this step is justified in 2 ways: The glycolytic process (up to this step) is now irreversible, and the energy supplied destabilizes the molecule. Because the reaction catalyzed by Phosphofructokinase 1 (PFK-1) is coupled to the hydrolysis of ATP, an energetically favorable step, it is, in essence, irreversible, and a different pathway must be used to do the reverse conversion during gluconeogenesis. This makes the reaction a key regulatory point (see below). This is also the rate-limiting step.

Furthermore, the second phosphorylation event is necessary to allow the formation of two charged groups (rather than only one) in the subsequent step of glycolysis, ensuring the prevention of free diffusion of substrates out of the cell.

The same reaction can also be catalyzed by pyrophosphate-dependent phosphofructokinase (**PFP** or **PPi-PFK**), which is found in most plants, some bacteria, archea, and protists, but not in animals. This enzyme uses pyrophosphate (PPi) as a phosphate donor instead of ATP. It is a reversible reaction, increasing the flexibility of glycolytic metabolism.[7] A rarer ADP-dependent PFK enzyme variant has been identified in archaean species.[8]

Cofactors: Mg^{2+}

β-D-Fructose 6-phosphate **(F6P)**	phosphofructokinase **(PFK-1)** *a transferase*	β-D-Fructose 1,6-bisphosphate **(F1,6BP)**

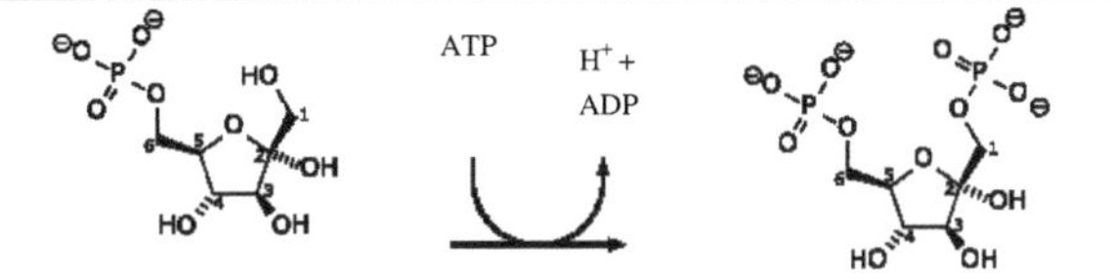

Destabilizing the molecule in the previous reaction allows the hexose ring to be split by aldolase into two triose sugars, dihydroxyacetone phosphate, a ketone, and glyceraldehyde 3-phosphate, an aldehyde. There are two classes of aldolases: class I aldolases, present in animals and plants, and class II aldolases, present in fungi and bacteria; the two classes use different mechanisms in cleaving the ketose ring.

Electrons delocalized in the carbon-carbon bond cleavage associate with the alcohol group. The resulting carbanion is stabilized by the structure of the carbanion itself via resonance charge distribution and by the presence of a charged ion prosthetic group.

β-D-Fructose 1,6-bisphosphate **(F1,6BP)**	fructose-bisphosphate aldolase (**ALDO**) *a lyase*	D-glyceraldehyde 3-phosphate **(GADP)**	Dihydroxyacetone phosphate (**DHAP**)

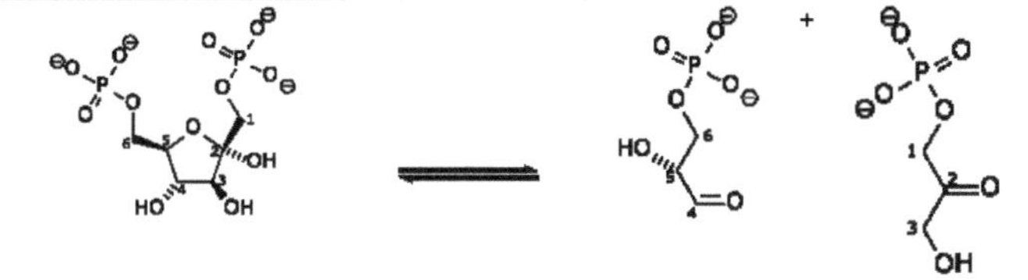

Triosephosphate isomerase rapidly interconverts dihydroxyacetone phosphate with glyceraldehyde 3-phosphate (**GADP**) that proceeds further into glycolysis. This is advantageous, as it directs dihydroxyacetone phosphate down the same pathway as glyceraldehyde 3-phosphate, simplifying regulation.

Dihydroxyacetone phosphate (**DHAP**)	triosephosphate isomerase (**TPI**) *an isomerase*	D-glyceraldehyde 3-phosphate (**GADP**)

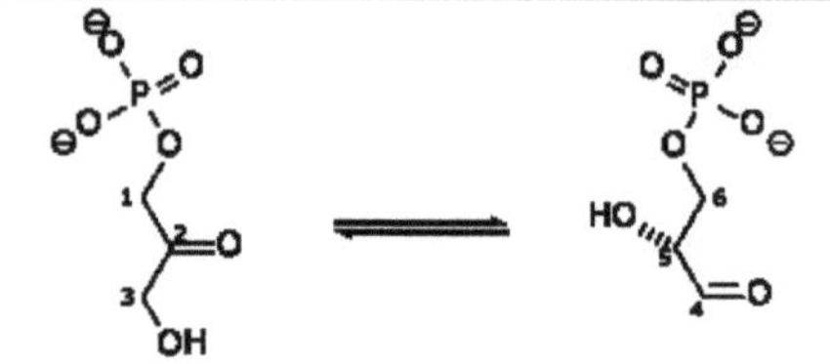

Pay-off phase

The second half of glycolysis is known as the pay-off phase, characterised by a net gain of the energy-rich molecules ATP and NADH.[5] Since glucose leads to two triose sugars in the preparatory phase, each reaction in the pay-off phase occurs twice per glucose molecule. This yields 2 NADH molecules and 4 ATP molecules, leading to a net gain of 2 NADH molecules and 2 ATP molecules from the glycolytic pathway per glucose.

The triose sugars are dehydrogenated and inorganic phosphate is added to them, forming 1,3-bisphosphoglycerate.

The hydrogen is used to reduce two molecules of NAD^+, a hydrogen carrier, to give NADH + H^+ for each triose.

Hydrogen atom balance and charge balance are both maintained because the phosphate (P_i) group actually exists in the form of a hydrogen phosphate anion (HPO_4^{2-}),[6] which dissociates to contribute the extra H^+ ion and gives a net charge of -3 on both sides.

glyceraldehyde 3-phosphate (**GADP**)	glyceraldehyde phosphate dehydrogenase (**GAPDH**) *an oxidoreductase*	D-1,3-bisphosphoglycerate (**1,3BPG**)

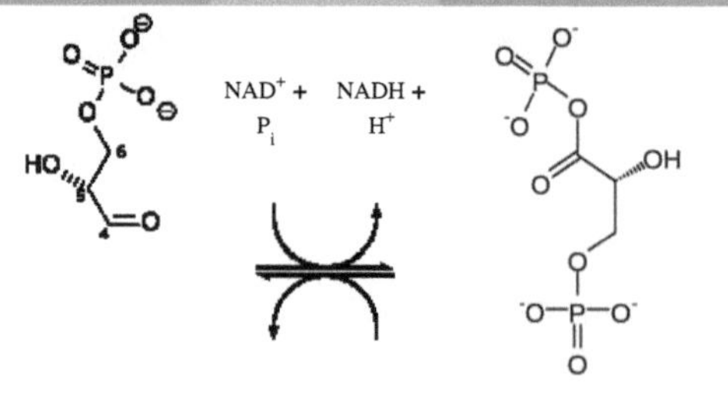

This step is the enzymatic transfer of a phosphate group from 1,3-bisphosphoglycerate to ADP by phosphoglycerate kinase, forming ATP and 3-phosphoglycerate. At this step, glycolysis has reached the break-even point: 2 molecules of ATP were consumed, and 2 new molecules have now been synthesized. This step, one of the two substrate-level phosphorylation steps, requires ADP; thus, when the cell has plenty of ATP (and little ADP), this reaction does not occur. Because ATP decays relatively quickly when it is not metabolized, this is an important regulatory point in the glycolytic pathway..

ADP actually exists as $ADPMg^-$, and ATP as $ATPMg^{2-}$, balancing the charges at -5 both sides.

Cofactors: Mg^{2+}

1,3-bisphosphoglycerate (**1,3-BPG**)	phosphoglycerate kinase (**PGK**) *a transferase*	3-phosphoglycerate (**3-P-G**)

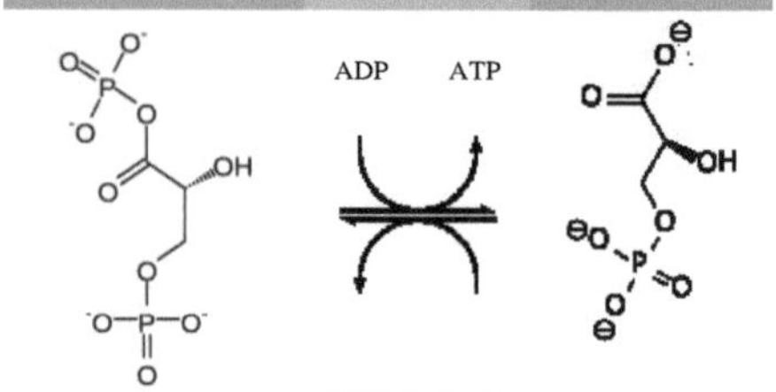

phosphoglycerate kinase (**PGK**)

Phosphoglycerate mutase now forms 2-phosphoglycerate.

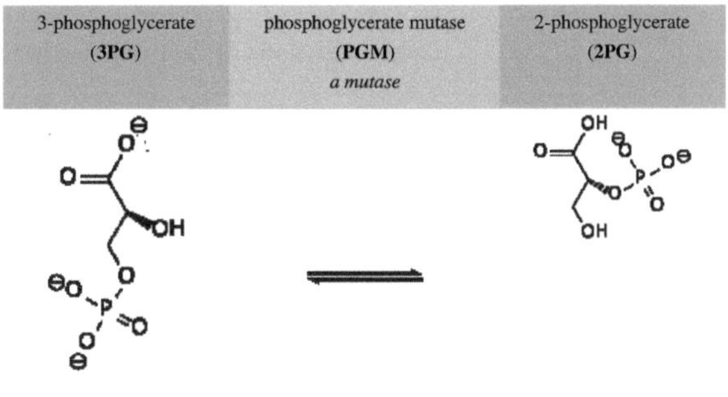

Enolase next forms phosphoenolpyruvate from 2-phosphoglycerate.

Cofactors: 2 Mg^{2+}: one "conformational" ion to coordinate with the carboxylate group of the substrate, and one "catalytic" ion that participates in the dehydration.

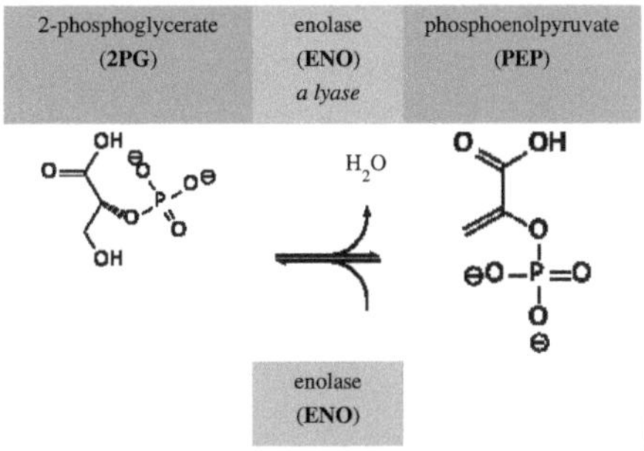

A final substrate-level phosphorylation now forms a molecule of pyruvate and a molecule of ATP by means of the enzyme pyruvate kinase. This serves as an additional regulatory step, similar to the phosphoglycerate kinase step.

Cofactors: Mg^{2+}

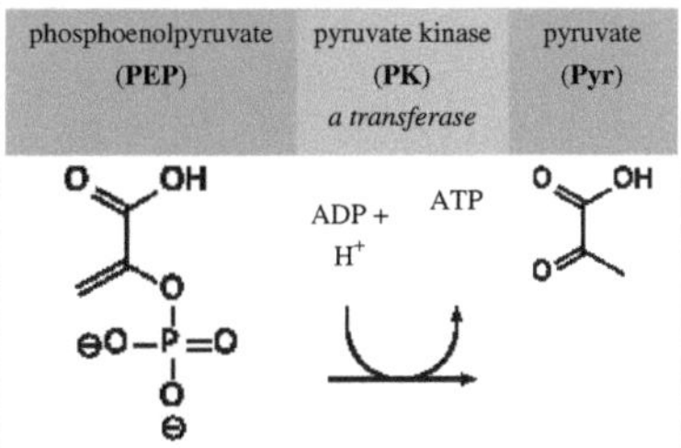

Regulation

Glycolysis is regulated by slowing down or speeding up certain steps in the glycolysis pathway. This is accomplished by inhibiting or activating the enzymes that are involved. The steps that are regulated may be determined by calculating the change in free energy, ΔG, for each step. If a step's products and reactants are in equilibrium, then the step is assumed not to be regulated. Since the change in free energy is zero for a system at equilibrium, *any step with a free energy change near zero is not being regulated.* If a step is being regulated, then that step's enzyme is not converting reactants into products as fast as it could, resulting in a build-up of reactants, which would be converted to products if the enzyme were operating faster. Since the reaction is thermodynamically favorable, the change in free energy for the step will be negative. *A step with a large negative change in free energy is assumed to be regulated.*

Free energy changes

Concentrations of metabolites in erythrocytes[9]

Compound	Concentration / m*M*
glucose	5.0
glucose-6-phosphate	0.083
fructose-6-phosphate	0.014
fructose-1,6-bisphosphate	0.031
dihydroxyacetone phosphate	0.14
glyceraldehyde-3-phosphate	0.019
1,3-bisphosphoglycerate	0.001
2,3-bisphosphoglycerate	4.0
3-phosphoglycerate	0.12
2-phosphoglycerate	0.03
phosphoenolpyruvate	0.023
pyruvate	0.051
ATP	1.85
ADP	0.14
P_i	1.0

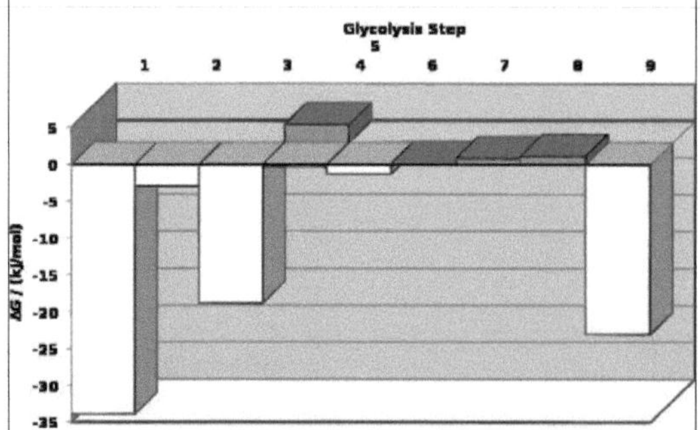

The change in free energy for each step of glycolysis estimated from the concentration of metabolites in an erythrocyte.

The change in free energy, ΔG, for each step in the glycolysis pathway can be calculated using $\Delta G = \Delta G^{\circ\prime} + RT\ln Q$, where Q is the reaction quotient. This requires knowing the concentrations of the metabolites. All of these values are available for erythrocytes, with the exception of the concentrations of NAD^+ and NADH. The ratio of NAD^+ to NADH in the cytoplasm is approximately 1000, which makes the oxidation of glyceraldehyde-3-phosphate (step 6) more favourable.

Using the measured concentrations of each step, and the standard free energy changes, the actual free energy change can be calculated. (Neglecting this is very common - the delta G of ATP hydrolysis in cells is not the standard free energy change of ATP hydrolysis quoted in textbooks).

Change in free energy for each step of glycolysis[10]

Step	Reaction	$\Delta G°'$ / (kJ/mol)	ΔG / (kJ/mol)
1	glucose + ATP^{4-} → glucose-6-phosphate^{2-} + ADP^{3-} + H^+	-16.7	-34
2	glucose-6-phosphate^{2-} → fructose-6-phosphate^{2-}	1.67	-2.9
3	fructose-6-phosphate^{2-} + ATP^{4-} → fructose-1,6-bisphosphate^{4-} + ADP^{3-} + H^+	-14.2	-19
4	fructose-1,6-bisphosphate^{4-} → dihydroxyacetone phosphate^{2-} + glyceraldehyde-3-phosphate^{2-}	23.9	-0.23
5	dihydroxyacetone phosphate^{2-} → glyceraldehyde-3-phosphate^{2-}	7.56	2.4
6	glyceraldehyde-3-phosphate^{2-} + P_i^{2-} + NAD^+ → 1,3-bisphosphoglycerate^{4-} + NADH + H^+	6.30	-1.29
7	1,3-bisphosphoglycerate^{4-} + ADP^{3-} → 3-phosphoglycerate^{3-} + ATP^{4-}	-18.9	0.09
8	3-phosphoglycerate^{3-} → 2-phosphoglycerate^{3-}	4.4	0.83
9	2-phosphoglycerate^{3-} → phosphoenolpyruvate^{3-} + H_2O	1.8	1.1
10	phosphoenolpyruvate^{3-} + ADP^{3-} + H^+ → pyruvate$^-$ + ATP^{4-}	-31.7	-23.0

From measuring the physiological concentrations of metabolites in an erythrocyte it seems that about seven of the steps in glycolysis are in equilibrium for that cell type. Three of the steps — the ones with large negative free energy changes — are not in equilibrium and are referred to as *irreversible*; such steps are often subject to regulation.

Step 5 in the figure is shown behind the other steps, because that step is a side-reaction that can decrease or increase the concentration of the intermediate glyceraldehyde-3-phosphate. That compound is converted to dihydroxyacetone phosphate by the enzyme triose phosphate isomerase, which is a catalytically perfect enzyme; its rate is so fast that the reaction can be assumed to be in equilibrium. The fact that ΔG is not zero indicates that the actual concentrations in the erythrocyte are not accurately known.

Biochemical logic

The existence of more than one point of regulation indicates that intermediates between those points enter and leave the glycolysis pathway by other processes. For example, in the first regulated step, hexokinase converts glucose into glucose-6-phosphate. Instead of continuing through the glycolysis pathway, this intermediate can be converted into glucose storage molecules, such as glycogen or starch. The reverse reaction, breaking down, e.g., glycogen, produces mainly glucose-6-phosphate; very little free glucose is formed in the reaction. The glucose-6-phosphate so produced can enter glycolysis *after* the first control point.

In the second regulated step (the third step of glycolysis), phosphofructokinase converts fructose-6-phosphate into fructose-1,6-bisphosphate, which then is converted into glyceraldehyde-3-phosphate and dihydroxyacetone phosphate. The dihydroxyacetone phosphate can be removed from glycolysis by conversion into glycerol-3-phosphate, which can be used to form triglycerides.[11] On the converse, triglycerides can be broken down into fatty acids and glycerol; the latter, in turn, can be converted into dihydroxyacetone phosphate, which can enter glycolysis *after* the second control point.

Regulation

The three regulated enzymes are hexokinase, phosphofructokinase, and pyruvate kinase.

The flux through the glycolytic pathway is adjusted in response to conditions both inside and outside the cell. The rate in liver is regulated to meet major cellular needs: (1) the production of ATP, (2) the provision of building blocks for biosynthetic reactions, and (3) to lower blood glucose, one of the major functions of the liver. When blood sugar falls, glycolysis is halted in the liver to allow the reverse process, gluconeogenesis. In glycolysis, the reactions catalyzed by hexokinase, phosphofructokinase, and pyruvate kinase are effectively irreversible in most organisms. In metabolic pathways, such enzymes are potential sites of control, and all three enzymes serve this purpose in glycolysis.

Hexokinase

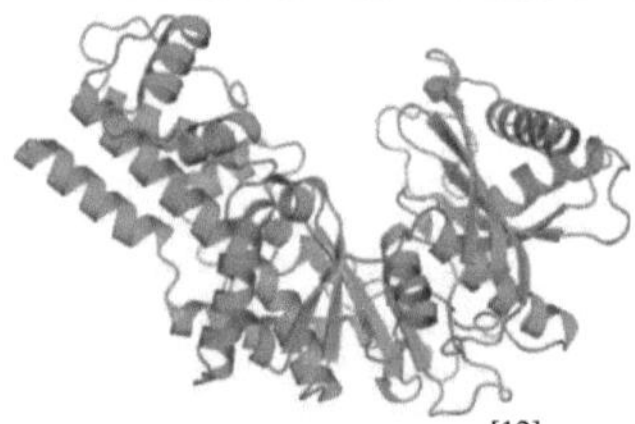
Yeast hexokinase B. PDB 1IG8 [12].

In animals, regulation of blood glucose levels by the pancreas in conjunction with the liver is a vital part of homeostasis. In liver cells, extra G6P (glucose-6-phosphate) may be converted to G1P for conversion to glycogen, or it is alternatively converted by glycolysis to acetyl-CoA and then citrate. Excess citrate is exported to the cytosol, where ATP citrate lyase will regenerate acetyl-CoA and OAA. The acetyl-CoA is then used for fatty acid synthesis and cholesterol synthesis, two important ways of utilizing excess glucose when its concentration is high in blood. Liver contains both hexokinase and glucokinase; both catalyse the phosphorylation of glucose to G6P but the latter is not inhibited by G6P. Thus, glucokinase allows glucose to be converted into glycogen, fatty acids, and cholesterol even as G6P accumulates in hepatocytes.[13] This is important when blood glucose levels are high. During hypoglycemia, the glycogen can be converted back to G6P and then converted to glucose by the liver-specific enzyme glucose 6-phosphatase and released into the blood without taking up the low concentration of glucose it releases. This reverse reaction is an important role of liver cells to maintain blood sugars levels during fasting. This is critical for brain function, since the brain utilizes glucose as an energy source under most conditions.

Phosphofructokinase

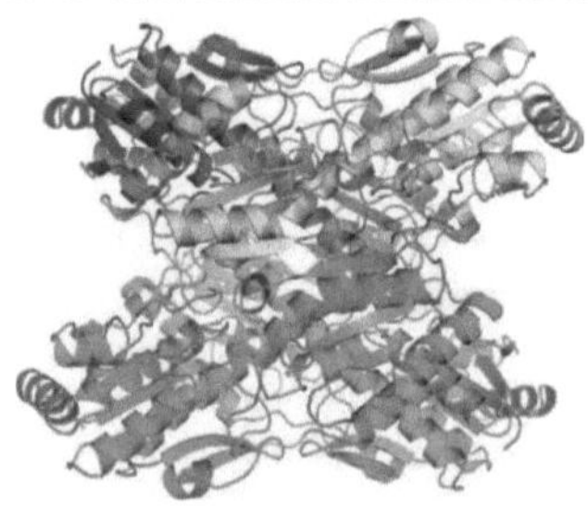
Bacillus stearothermophilus phosphofructokinase. PDB 6PFK [14].

Phosphofructokinase is an important control point in the glycolytic pathway, since it is one of the irreversible steps and has key allosteric effectors, AMP and fructose 2,6-bisphosphate (F2,6BP).

Fructose 2,6-bisphosphate (F2,6BP) is a very potent activator of phosphofructokinase (PFK-1), which is synthesized when F6P is phosphorylated by a second phosphofructokinase (PFK2). In liver, when blood sugar is low and glucagon elevates cAMP, PFK2 is phosphorylated by protein kinase A. The phosphorylation inactivates PFK2, and another domain on this protein becomes active as fructose 2,6-bisphosphatase, which converts F2,6BP back to F6P. Both glucagon and epinephrine cause high levels of cAMP in the liver. The result of lower levels of liver fructose-2,6-bisphosphate is a decrease in activity of phosphofructokinase and an increase in activity of fructose 1,6-bisphosphatase, so that gluconeogenesis (in essence, "glycolysis in reverse") is favored. This is consistent with the role of the liver in such situations, since the response of the liver to these hormones is to release glucose to the blood.

ATP competes with AMP for the allosteric effector site on the PFK enzyme. ATP concentrations in cells are much higher than those of AMP, typically 100-fold higher,[15] but the concentration of ATP does not change more than about 10% under physiological conditions, whereas a 10% drop in ATP results in a 6-fold increase in AMP.[16] Thus, the relevance of ATP as an allosteric effector is questionable. An increase in AMP is a consequence of a decrease in energy charge in the cell.

Citrate inhibits phosphofructokinase when tested *in vitro* by enhancing the inhibitory effect of ATP. However, it is doubtful that this is a meaningful effect *in vivo*, because citrate in the cytosol is utilized mainly for conversion to acetyl-CoA for fatty acid and cholesterol synthesis.

Pyruvate kinase

This enzyme catalyzes the last step of glycolysis, in which pyruvate and ATP are formed. Regulation of this enzyme is discussed in the main topic, pyruvate kinase.

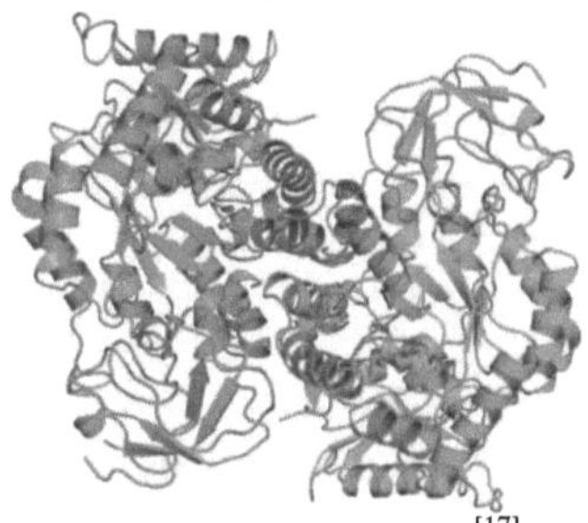

Yeast pyruvate kinase. PDB 1A3W [17].

Post-glycolysis processes

The overall process of glycolysis is:

$$\text{glucose} + 2\ \text{NAD}^+ + 2\ \text{ADP} + 2\ \text{P}_i \rightarrow 2\ \text{pyruvate} + 2\ \text{NADH} + 2\ \text{H}^+ + 2\ \text{ATP} + 2\ \text{H}_2\text{O}$$

If glycolysis were to continue indefinitely, all of the NAD^+ would be used up, and glycolysis would stop. To allow glycolysis to continue, organisms must be able to oxidize NADH back to NAD^+.

Fermentation

One method of doing this is to simply have the pyruvate do the oxidation; in this process, the pyruvate is converted to lactate (the conjugate base of lactic acid) in a process called lactic acid fermentation:

$$\text{pyruvate} + \text{NADH} + \text{H}^+ \rightarrow \text{lactate} + \text{NAD}^+$$

This process occurs in the bacteria involved in making yogurt (the lactic acid causes the milk to curdle). This process also occurs in animals under hypoxic (or partially anaerobic) conditions, found, for example, in overworked muscles that are starved of oxygen, or in infarcted heart muscle cells. In many tissues, this is a cellular last resort for energy; most animal tissue cannot tolerate anaerobic conditions for an extended period of time.

Some organisms, such as yeast, convert NADH back to NAD^+ in a process called ethanol fermentation. In this process, the pyruvate is converted first to acetaldehyde and carbon dioxide, then to ethanol.

Lactic acid fermentation and ethanol fermentation can occur in the absence of oxygen. This anaerobic fermentation allows many single-cell organisms to use glycolysis as their only energy source.

Anaerobic respiration

In the above two examples of fermentation, NADH is oxidized by transferring two electrons to pyruvate. However, anaerobic bacteria use a wide variety of compounds as the terminal electron acceptors in cellular respiration: nitrogenous compounds, such as nitrates and nitrites; sulfur compounds, such as sulfates, sulfites, sulfur dioxide, and elemental sulfur; carbon dioxide; iron compounds; manganese compounds; cobalt compounds; and uranium compounds.

Aerobic respiration

In aerobic organisms, a complex mechanism has been developed to use the oxygen in air as the final electron acceptor.

- First, pyruvate is converted to acetyl-CoA and CO_2 within the mitochondria in a process called pyruvate decarboxylation.
- Second, the acetyl-CoA enters the citric acid cycle, also known as Krebs Cycle, where it is fully oxidized to carbon dioxide and water, producing yet more NADH.
- Third, the NADH is oxidized to NAD^+ by the electron transport chain, using oxygen as the final electron acceptor. This process creates a hydrogen ion gradient across the inner membrane of the mitochondria.
- Fourth, the proton gradient is used to produce about 2.5 ATP for every NADH oxidized in a process called oxidative phosphorylation.

Intermediates for other pathways

This article concentrates on the catabolic role of glycolysis with regard to converting potential chemical energy to usable chemical energy during the oxidation of glucose to pyruvate. Many of the metabolites in the glycolytic pathway are also used by anabolic pathways, and, as a consequence, flux through the pathway is critical to maintain a supply of carbon skeletons for biosynthesis.

In addition, not all carbon entering the pathway leaves as pyruvate and may be extracted at earlier stages to provide carbon compounds for other pathways.

These metabolic pathways are all strongly reliant on glycolysis as a source of metabolites: and many more.

- Gluconeogenesis
- Lipid metabolism
- Pentose phosphate pathway
- Citric acid cycle, which in turn leads to:
  - Amino acid synthesis
  - Nucleotide synthesis
  - Tetrapyrrole synthesis

From an anabolic metabolism perspective, the NADH has a role to drive synthetic reactions, doing so by directly or indirectly reducing the pool of NADP+ in the cell to NADPH, which is another important reducing agent for biosynthetic pathways in a cell.

Glycolysis in disease

Genetic diseases

Glycolytic mutations are generally rare due to importance of the metabolic pathway, this means that the majority of occurring mutations result in an inability for the cell to respire, and therefore cause the death of the cell at an early stage. However, some mutations are seen with one notable example being Pyruvate kinase deficiency, leading to chronic hemolytic anemia.

Cancer

Malignant rapidly growing tumor cells typically have glycolytic rates that are up to 200 times higher than those of their normal tissues of origin. This phenomenon was first described in 1930 by Otto Warburg and is referred to as the Warburg effect. The Warburg hypothesis claims that cancer is primarily caused by dysfunctionality in mitochondrial metabolism, rather than because of uncontrolled growth of cells. A number of theories have been advanced to explain the Warburg effect. One such theory suggests that the increased glycolysis is a normal protective process of the body and that malignant change could be primarily caused by energy metabolism.[18]

This high glycolysis rate has important medical applications, as high aerobic glycolysis by malignant tumors is utilized clinically to diagnose and monitor treatment responses of cancers by imaging uptake of 2-^{18}F-2-deoxyglucose (FDG) (a radioactive modified hexokinase substrate) with positron emission tomography (PET).[19] [20]

There is ongoing research to affect mitochondrial metabolism and treat cancer by reducing glycolysis and thus starving cancerous cells in various new ways, including a ketogenic diet.[21]

Alzheimer's disease

Disfunctioning glycolysis or glucose metabolism in fronto-temporo-parietal and cingulate cortices has been associated with Alzheimer's disease,[] probably due to the decreased amyloid β (1-42) (Aβ42) and increased tau, phosphorylated tau in cerebrospinal fluid (CSF)[] These changes in CSF biomarkers are indicative that a patient has Alzheimer's disease

Interactive pathway map

Click on genes, proteins and metabolites below to link to respective articles. [22]

[[File:

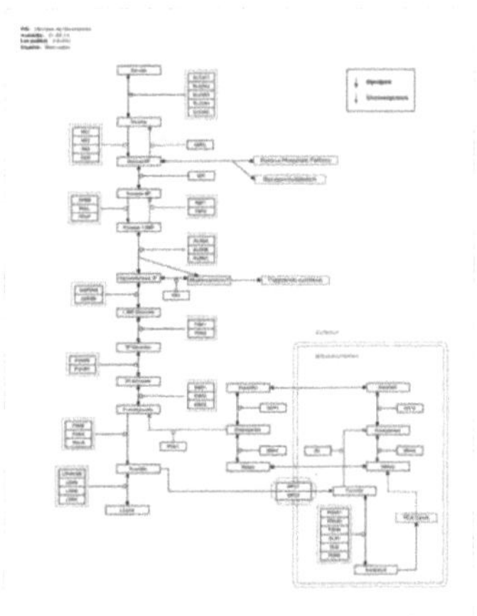

{{{bSize}}}px

Glycolysis and Gluconeogenesis edit [23]

[1] Webster's New International Dictionary of the English Language, 2nd ed. (1937) Merriam Company, Springfield, Mass.
[3] Romano AH, Conway T. (1996) Evolution of carbohydrate metabolic pathways. *Res Microbiol.* 147(6–7):448–55 PMID 9084754
[4] Kim BH, Gadd GM. (2011) Bacterial Physiology and Metabolism, 3rd edition.
[5] Glycolysis – Animation and Notes (http://pharmaxchange.info/press/2011/09/glycolysis-animation-and-notes/)
[6] Error: Bad DOI specified: 10.1007/s11306-008-0142-2
[12] http://www.rcsb.org/pdb/explore/explore.do?structureId=1IG8
[13] Voet D., and Voet J. G. (2004). Biochemistry 3rd Edition (New York, John Wiley & Sons, Inc.)
[14] http://www.rcsb.org/pdb/explore/explore.do?structureId=6PFK
[15] Beis I., and Newsholme E. A. (1975). The contents of adenine nucleotides, phosphagens and some glycolytic intermediates in resting muscles from vertebrates and invertebrates. Biochem J 152, 23-32.
[16] Voet D., and Voet J. G. (2004). Biochemistry 3rd Edition (New York, John Wiley & Sons, Inc.).
[17] http://www.rcsb.org/pdb/explore/explore.do?structureId=1A3W
[22] The interactive pathway map can be edited at WikiPathways:
[23] http://www.wikipathways.org/index.php/Pathway:WP534

Alternative nomenclature

Some of the metabolites in glycolysis have alternative names and nomenclature. In part, this is because some of them are common to other pathways, such as the Calvin cycle.

	This article		Alternative names	Alternative nomenclature
1	glucose	**Glc**	dextrose	
3	fructose 6-phosphate	**F6P**		
4	fructose 1,6-bisphosphate	**F1,6BP**	fructose 1,6-diphosphate	**FBP**, **FDP**, **F1,6DP**
5	dihydroxyacetone phosphate	**DHAP**	glycerone phosphate	
6	glyceraldehyde 3-phosphate	**GADP**	3-phosphoglyceraldehyde	**PGAL**, **G3P**, **GALP**,**GAP**,**TP**
7	1,3-bisphosphoglycerate	**1,3BPG**	glycerate 1,3-bisphosphate, glycerate 1,3-diphosphate, 1,3-diphosphoglycerate	**PGAP**, **BPG**, **DPG**
8	3-phosphoglycerate	**3PG**	glycerate 3-phosphate	**PGA**, **GP**
9	2-phosphoglycerate	**2PG**	glycerate 2-phosphate	
10	phosphoenolpyruvate	**PEP**		
11	pyruvate	**Pyr**	pyruvic acid	

See also

- carbohydrate catabolism
- Citric acid cycle
- Cori cycle
- Fermentation (biochemistry)
- Gluconeogenesis
- Pentose phosphate pathway
- Pyruvate decarboxylation
- Triose kinase

References

External links

- A Detailed Glycolysis Animation provided by [[IUBMB (http://www.iubmb-nicholson.org/swf/glycolysis.swf)]] (Adobe Flash (http://get.adobe.com/flashplayer/) Required)
- The Glycolytic enzymes in Glycolysis (http://nist.rcsb.org/pdb/molecules/pdb50_1.html) at RCSB PDB
- Glycolytic cycle with animations (http://www.wdv.com/CellWorld/Biochemistry/Glycolytic) at wdv.com
- Metabolism, Cellular Respiration and Photosynthesis - The Virtual Library of Biochemistry and Cell Biology (http://www.biochemweb.org/metabolism.shtml) at biochemweb.org
- notes on glycolysis (http://www.rahulgladwin.com/blog/2007/01/notes-on-glycolysis.html) at rahulgladwin.com
- The chemical logic behind glycolysis (http://www2.ufp.pt/~pedros/bq/glycolysis.htm) at ufp.pt
- Expasy biochemical pathways poster (http://www.expasy.org/tools/pathways/boehringer_legends.html) at ExPASy
- Mnemonic at medicalmnemonics.com *317 5468* (http://www.medicalmnemonics.com/cgi-bin/lookup.cfm?id1=317&id2=5468&id3=&id4=)

Oxidative_phosphorylation

Oxidative phosphorylation (or OXPHOS in short) is the metabolic pathway in which the mitochondria in cells use their structure, enzymes, and energy released by the oxidation of nutrients to reform ATP. Although the many forms of life on earth use a range of different nutrients, ATP is the molecule that supplies energy to metabolism. Almost all aerobic organisms carry out oxidative phosphorylation. This pathway is probably so pervasive because it is a highly efficient way of releasing energy, compared to alternative fermentation processes such as anaerobic glycolysis.

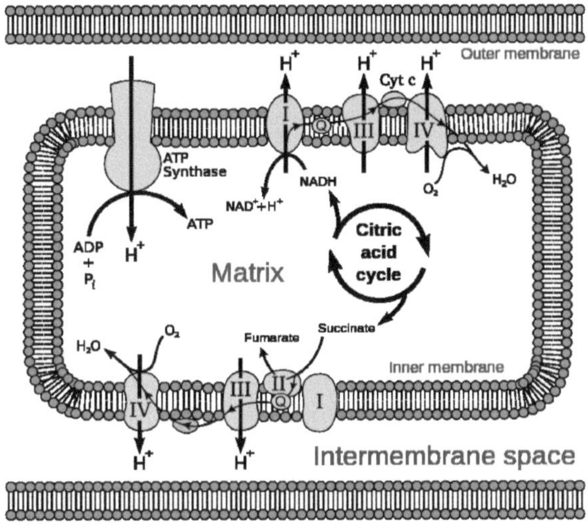

The electron transport chain in the mitochondrion is the site of oxidative phosphorylation in eukaryotes. The NADH and succinate generated in the citric acid cycle are oxidized, releasing energy to power the ATP synthase.

During oxidative phosphorylation, electrons are transferred from electron donors to electron acceptors such as oxygen, in redox reactions. These redox reactions release energy, which is used to form ATP. In eukaryotes, these redox reactions are carried out by a series of protein complexes within the cell's intermembrane wall mitochondria, whereas, in prokaryotes, these proteins are located in the cells' intermembrane space. These linked sets of proteins are called electron transport chains. In eukaryotes, five main protein complexes are involved, whereas in prokaryotes many different enzymes are present, using a variety of electron donors and acceptors.

The energy released by electrons flowing through this electron transport chain is used to transport protons across the inner mitochondrial membrane, in a process called *electron transport*. This generates potential energy in the form of a pH gradient and an electrical potential across this membrane. This store of energy is tapped by allowing protons to flow back across the membrane and down this gradient, through a large enzyme called ATP synthase; this process is known as chemiosmosis. This enzyme uses this energy to generate ATP from adenosine diphosphate (ADP), in a phosphorylation reaction. This reaction is driven by the proton flow, which forces the rotation of a part of the enzyme; the ATP synthase is a rotary mechanical motor.

Although oxidative phosphorylation is a vital part of metabolism, it produces reactive oxygen species such as superoxide and hydrogen peroxide, which lead to propagation of free radicals, damaging cells and contributing to disease and, possibly, aging (senescence). The enzymes carrying out this metabolic pathway are also the target of many drugs and poisons that inhibit their activities.

Overview of energy transfer by chemiosmosis

Oxidative phosphorylation works by using energy-releasing chemical reactions to drive energy-requiring reactions: The two sets of reactions are said to be *coupled*. This means one cannot occur without the other. The flow of electrons through the electron transport chain, from electron donors such as NADH to electron acceptors such as oxygen, is an exergonic process – it releases energy, whereas the synthesis of ATP is an endergonic process, which requires an input of energy. Both the electron transport chain and the ATP synthase are embedded in a membrane, and energy is transferred from electron transport chain to the ATP synthase by movements of protons across this membrane, in a process called *chemiosmosis*.[1] In practice, this is like a simple electric circuit, with a current of protons being driven from the negative N-side of the membrane to the positive P-side by the proton-pumping enzymes of the electron transport chain. These enzymes are like a battery, as they perform work to drive current through the circuit. The movement of protons creates an electrochemical gradient across the membrane, which is often called the *proton-motive force*. It has two components: a difference in proton concentration (a H+ gradient, ΔpH) and a difference in electric potential, with the N-side having a negative charge.[]

ATP synthase releases this stored energy by completing the circuit and allowing protons to flow down the electrochemical gradient, back to the N-side of the membrane.[] This kinetic energy drives the rotation of part of the enzymes structure and couples this motion to the synthesis of ATP.

The two components of the proton-motive force are thermodynamically equivalent: In mitochondria, the largest part of energy is provided by the potential; in alkaliphile bacteria the electrical energy even has to compensate for a counteracting inverse pH difference. Inversely, chloroplasts operate mainly on ΔpH. However, they also require a small membrane potential for the kinetics of ATP synthesis. At least in the case of the fusobacterium *P. modestum* it drives the counter-rotation of subunits a and c of the F_O motor of ATP synthase.[]

The amount of energy released by oxidative phosphorylation is high, compared with the amount produced by anaerobic fermentation. Glycolysis produces only 2 ATP molecules, but somewhere between 30 and 36 ATPs are produced by the oxidative phosphorylation of the 10 NADH and 2 succinate molecules made by converting one molecule of glucose to carbon dioxide and water,[2] while each cycle of beta oxidation of a fatty acid yields about 14 ATPs. These ATP yields are theoretical maximum values; in practice, some protons leak across the membrane, lowering the yield of ATP.[3]

Electron and proton transfer molecules

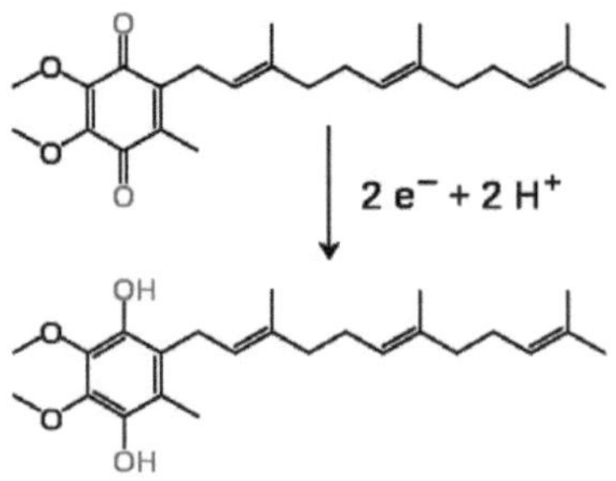

Reduction of coenzyme Q from its ubiquinone form (Q) to the reduced ubiquinol form (QH_2).

The electron transport chain carries both protons and electrons, passing electrons from donors to acceptors, and transporting protons across a membrane. These processes use both soluble and protein-bound transfer molecules. In mitochondria, electrons are transferred within the intermembrane space by the water-soluble electron transfer protein cytochrome c.[4] This carries only electrons, and these are transferred by the reduction and oxidation of an iron atom that the protein holds within a heme group in its structure. Cytochrome c is also found in some bacteria, where it is located within the periplasmic space.[5]

Within the inner mitochondrial membrane, the lipid-soluble electron carrier coenzyme Q10 (Q) carries both electrons and protons by a redox cycle.[6] This small benzoquinone molecule is very hydrophobic, so it diffuses freely within the membrane. When Q accepts two electrons and two protons, it becomes reduced to the *ubiquinol* form (QH_2); when QH_2 releases two electrons and two protons, it becomes oxidized back to the *ubiquinone* (Q) form. As a result, if two enzymes are arranged so that Q is reduced on one side of the membrane and QH_2 oxidized on the other, ubiquinone will couple these reactions and shuttle protons across the membrane.[7] Some bacterial electron transport chains use different quinones, such as menaquinone, in addition to ubiquinone.[]

Within proteins, electrons are transferred between flavin cofactors,[] [] iron–sulfur clusters, and cytochromes. There are several types of iron–sulfur cluster. The simplest kind found in the electron transfer chain consists of two iron atoms joined by two atoms of inorganic sulfur; these are called [2Fe–2S] clusters. The second kind, called [4Fe–4S], contains a cube of four iron atoms and four sulfur atoms. Each iron atom in these clusters is coordinated by an additional amino acid, usually by the sulfur atom of cysteine. Metal ion cofactors undergo redox reactions without binding or releasing protons, so in the electron transport chain they serve solely to transport electrons through proteins. Electrons move quite long distances through proteins by hopping along chains of these cofactors.[8] This occurs by quantum tunnelling, which is rapid over distances of less than 1.4×10^{-9} m.[9]

Eukaryotic electron transport chains

Many catabolic biochemical processes, such as glycolysis, the citric acid cycle, and beta oxidation, produce the reduced coenzyme NADH. This coenzyme contains electrons that have a high transfer potential; in other words, they will release a large amount of energy upon oxidation. However, the cell does not release this energy all at once, as this would be an uncontrollable reaction. Instead, the electrons are removed from NADH and passed to oxygen through a series of enzymes that each release a small amount of the energy. This set of enzymes, consisting of complexes I through IV, is called the electron transport chain and is found in the inner membrane of the mitochondrion. Succinate is also oxidized by the electron transport chain, but feeds into the pathway at a different point.

In eukaryotes, the enzymes in this electron transport system use the energy released from the oxidation of NADH to pump protons across the inner membrane of the mitochondrion. This causes protons to build up in the intermembrane space, and generates an electrochemical gradient across the membrane. The energy stored in this potential is then used by ATP synthase to produce ATP. Oxidative phosphorylation in the eukaryotic mitochondrion is the best-understood example of this process. The mitochondrion is present in almost all eukaryotes, with the exception of anaerobic protozoa such as *Trichomonas vaginalis* that instead reduce protons to hydrogen in a remnant mitochondrion called a hydrogenosome.[10]

Typical respiratory enzymes and substrates in eukaryotes.

Respiratory enzyme	Redox pair	Midpoint potential (Volts)
NADH dehydrogenase	NAD^+ / NADH	−0.32[11]
Succinate dehydrogenase	FMN or FAD / $FMNH_2$ or $FADH_2$	−0.20[11]
Cytochrome bc_1 complex	Coenzyme $Q10_{ox}$ / Coenzyme $Q10_{red}$	+0.06[11]
Cytochrome bc_1 complex	Cytochrome b_{ox} / Cytochrome b_{red}	+0.12[11]
Complex IV	Cytochrome c_{ox} / Cytochrome c_{red}	+0.22[11]
Complex IV	Cytochrome a_{ox} / Cytochrome a_{red}	+0.29[11]
Complex IV	O_2 / HO^-	+0.82[11]
Conditions: pH = 7[11]		

NADH-coenzyme Q oxidoreductase (complex I)

NADH-coenzyme Q oxidoreductase, also known as *NADH dehydrogenase* or *complex I*, is the first protein in the electron transport chain.[] Complex I is a giant enzyme with the mammalian complex I having 46 subunits and a molecular mass of about 1,000 kilodaltons (kDa).[] The structure is known in detail only from a bacterium;[] [12] in most organisms the complex resembles a boot with a large "ball" poking out from the membrane into the mitochondrion.[13] [14] The genes that encode the individual proteins are contained in both the cell nucleus and the mitochondrial genome, as is the case for many enzymes present in the mitochondrion.

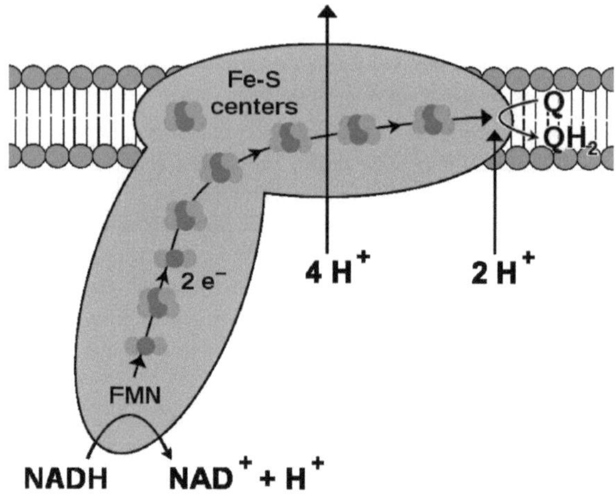

Complex I or NADH-Q oxidoreductase. The abbreviations are discussed in the text. In all diagrams of respiratory complexes in this article, the matrix is at the bottom, with the intermembrane space above.

The reaction that is catalyzed by this enzyme is the two electron oxidation of NADH by coenzyme Q10 or *ubiquinone* (represented as Q in the equation below), a lipid-soluble quinone that is found in the mitochondrion membrane:

The start of the reaction, and indeed of the entire electron chain, is the binding of a NADH molecule to complex I and the donation of two electrons. The electrons enter complex I via a prosthetic group attached to the complex, flavin mononucleotide (FMN). The addition of electrons to FMN converts it to its reduced form, $FMNH_2$. The electrons are then transferred through a series of iron–sulfur clusters: the second kind of prosthetic group present in the complex.[] There are both [2Fe–2S] and [4Fe–4S] iron–sulfur clusters in complex I.

As the electrons pass through this complex, four protons are pumped from the matrix into the intermembrane space. Exactly how this occurs is unclear, but it seems to involve conformational changes in complex I that cause the protein to bind protons on the N-side of the membrane and release them on the P-side of the membrane.[15] Finally, the electrons are transferred from the chain of iron–sulfur clusters to a ubiquinone molecule in the membrane.[] Reduction of ubiquinone also contributes to the generation of a proton gradient, as two protons are taken up from the matrix as it is reduced to ubiquinol (QH_2).

Succinate-Q oxidoreductase (complex II)

Succinate-Q oxidoreductase, also known as *complex II* or *succinate dehydrogenase*, is a second entry point to the electron transport chain.[16] It is unusual because it is the only enzyme that is part of both the citric acid cycle and the electron transport chain. Complex II consists of four protein subunits and contains a bound flavin adenine dinucleotide (FAD) cofactor, iron–sulfur clusters, and a heme group that does not participate in electron transfer to coenzyme Q, but is believed to be important in decreasing production of reactive oxygen species.[17] [18] It oxidizes succinate to fumarate and reduces ubiquinone. As this reaction releases less energy than the oxidation of NADH, complex II does not transport protons across the membrane and does not contribute to the proton gradient.

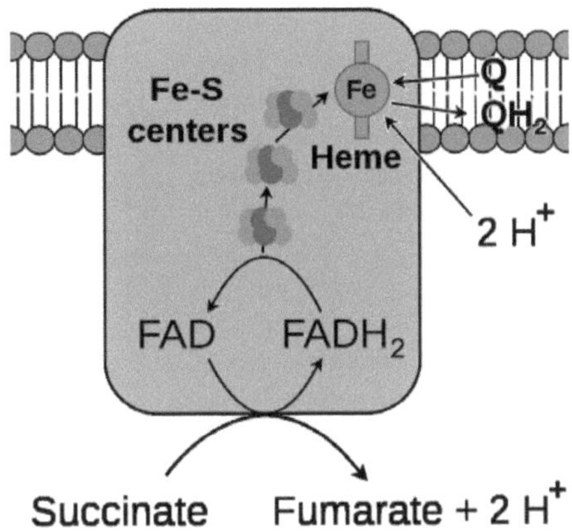

Complex II: Succinate-Q oxidoreductase.

In some eukaryotes, such as the parasitic worm *Ascaris suum*, an enzyme similar to complex II, fumarate reductase (menaquinol:fumarate oxidoreductase, or QFR), operates in reverse to oxidize ubiquinol and reduce fumarate. This allows the worm to survive in the anaerobic environment of the large intestine, carrying out anaerobic oxidative phosphorylation with fumarate as the electron acceptor.[19] Another unconventional function of complex II is seen in the malaria parasite *Plasmodium falciparum*. Here, the reversed action of complex II as an oxidase is important in regenerating ubiquinol, which the parasite uses in an unusual form of pyrimidine biosynthesis.[20]

Electron transfer flavoprotein-Q oxidoreductase

Electron transfer flavoprotein-ubiquinone oxidoreductase (ETF-Q oxidoreductase), also known as *electron transferring-flavoprotein dehydrogenase*, is a third entry point to the electron transport chain. It is an enzyme that accepts electrons from electron-transferring flavoprotein in the mitochondrial matrix, and uses these electrons to reduce ubiquinone.[21] This enzyme contains a flavin and a [4Fe–4S] cluster, but, unlike the other respiratory complexes, it attaches to the surface of the membrane and does not cross the lipid bilayer.[22]

In mammals, this metabolic pathway is important in beta oxidation of fatty acids and catabolism of amino acids and choline, as it accepts electrons from multiple acetyl-CoA dehydrogenases.[23] [24] In plants, ETF-Q oxidoreductase is also important in the metabolic responses that allow survival in extended periods of darkness.[25]

Q-cytochrome c oxidoreductase (complex III)

Q-cytochrome c oxidoreductase is also known as *cytochrome c reductase*, *cytochrome bc$_1$ complex*, or simply *complex III*.[26] [27] In mammals, this enzyme is a dimer, with each subunit complex containing 11 protein subunits, an [2Fe-2S] iron–sulfur cluster and three cytochromes: one cytochrome c_1 and two b cytochromes.[28] A cytochrome is a kind of electron-transferring protein that contains at least one heme group. The iron atoms inside complex III's heme groups alternate between a reduced ferrous (+2) and oxidized ferric (+3) state as the electrons are transferred through the protein.

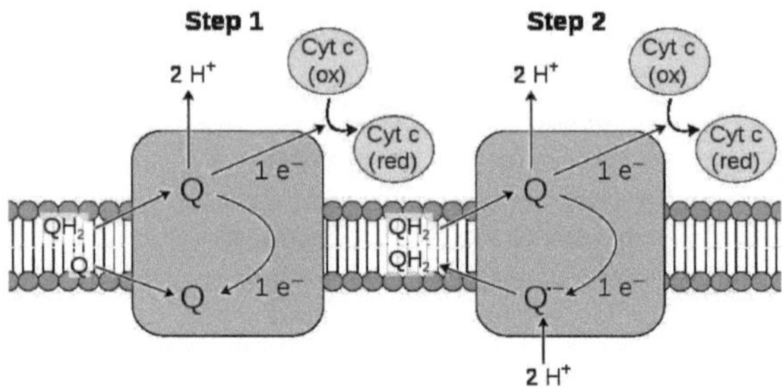

The two electron transfer steps in complex III: Q-cytochrome c oxidoreductase. After each step, Q (in the upper part of the figure) leaves the enzyme.

The reaction catalyzed by complex III is the oxidation of one molecule of ubiquinol and the reduction of two molecules of cytochrome c, a heme protein loosely associated with the mitochondrion. Unlike coenzyme Q, which carries two electrons, cytochrome c carries only one electron.

As only one of the electrons can be transferred from the QH_2 donor to a cytochrome c acceptor at a time, the reaction mechanism of complex III is more elaborate than those of the other respiratory complexes, and occurs in two steps called the Q cycle.[29] In the first step, the enzyme binds three substrates, first, QH_2, which is then oxidized, with one electron being passed to the second substrate, cytochrome c. The two protons released from QH_2 pass into the intermembrane space. The third substrate is Q, which accepts the second electron from the QH_2 and is reduced to $Q^{\cdot-}$, which is the ubisemiquinone free radical. The first two substrates are released, but this ubisemiquinone intermediate remains bound. In the second step, a second molecule of QH_2 is bound and again passes its first electron to a cytochrome c acceptor. The second electron is passed to the bound ubisemiquinone, reducing it to QH_2 as it gains two protons from the mitochondrial matrix. This QH_2 is then released from the enzyme.[30]

As coenzyme Q is reduced to ubiquinol on the inner side of the membrane and oxidized to ubiquinone on the other, a net transfer of protons across the membrane occurs, adding to the proton gradient.[] The rather complex two-step mechanism by which this occurs is important, as it increases the efficiency of proton transfer. If, instead of the Q cycle, one molecule of QH_2 were used to directly reduce two molecules of cytochrome c, the efficiency would be halved, with only one proton transferred per cytochrome c reduced.[]

Cytochrome c oxidase (complex IV)

Cytochrome c oxidase, also known as *complex IV*, is the final protein complex in the electron transport chain.[31] The mammalian enzyme has an extremely complicated structure and contains 13 subunits, two heme groups, as well as multiple metal ion cofactors – in all three atoms of copper, one of magnesium and one of zinc.[32]

This enzyme mediates the final reaction in the electron transport chain and transfers electrons to oxygen, while pumping protons across the membrane.[33] The final electron acceptor oxygen, which is also called the *terminal electron acceptor*, is reduced to water in this step. Both the direct pumping of protons and the consumption of matrix protons in the reduction of oxygen contribute to the proton gradient. The reaction catalyzed is the oxidation of cytochrome c and the reduction of oxygen:

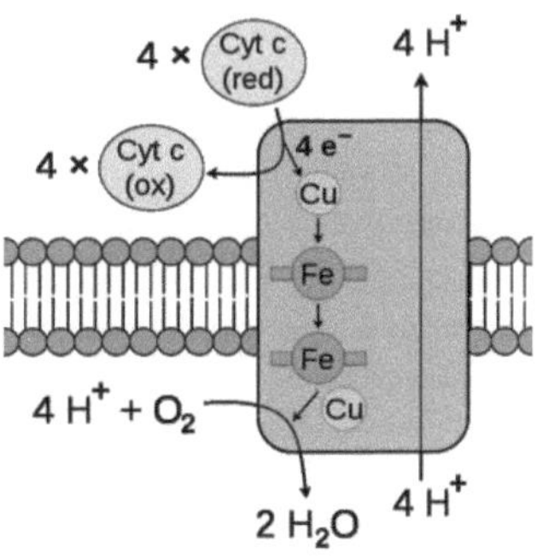

Complex IV: cytochrome c oxidase.

Alternative reductases and oxidases

Many eukaryotic organisms have electron transport chains that differ from the much-studied mammalian enzymes described above. For example, plants have alternative NADH oxidases, which oxidize NADH in the cytosol rather than in the mitochondrial matrix, and pass these electrons to the ubiquinone pool.[34] These enzymes do not transport protons, and, therefore, reduce ubiquinone without altering the electrochemical gradient across the inner membrane.[35]

Another example of a divergent electron transport chain is the *alternative oxidase*, which is found in plants, as well as some fungi, protists, and possibly some animals.[36] [37] This enzyme transfers electrons directly from ubiquinol to oxygen.[38]

The electron transport pathways produced by these alternative NADH and ubiquinone oxidases have lower ATP yields than the full pathway. The advantages produced by a shortened pathway are not entirely clear. However, the alternative oxidase is produced in response to stresses such as cold, reactive oxygen species, and infection by pathogens, as well as other factors that inhibit the full electron transport chain.[39] [40] Alternative pathways might, therefore, enhance an organisms' resistance to injury, by reducing oxidative stress.[41]

Organization of complexes

The original model for how the respiratory chain complexes are organized was that they diffuse freely and independently in the mitochondrial membrane.[] However, recent data suggest that the complexes might form higher-order structures called supercomplexes or "respirasomes."[42] In this model, the various complexes exist as organized sets of interacting enzymes.[43] These associations might allow channeling of substrates between the various enzyme complexes, increasing the rate and efficiency of electron transfer.[44] Within such mammalian supercomplexes, some components would be present in higher amounts than others, with some data suggesting a ratio between complexes I/II/III/IV and the ATP synthase of approximately 1:1:3:7:4.[45] However, the debate over this supercomplex hypothesis is not completely resolved, as some data do not appear to fit with this model.[] [46]

Prokaryotic electron transport chains

In contrast to the general similarity in structure and function of the electron transport chains in eukaryotes, bacteria and archaea possess a large variety of electron-transfer enzymes. These use an equally wide set of chemicals as substrates.[47] In common with eukaryotes, prokaryotic electron transport uses the energy released from the oxidation of a substrate to pump ions across a membrane and generate an electrochemical gradient. In the bacteria, oxidative phosphorylation in *Escherichia coli* is understood in most detail, while archaeal systems are at present poorly understood.[48]

The main difference between eukaryotic and prokaryotic oxidative phosphorylation is that bacteria and archaea use many different substances to donate or accept electrons. This allows prokaryotes to grow under a wide variety of environmental conditions.[] In *E. coli*, for example, oxidative phosphorylation can be driven by a large number of pairs of reducing agents and oxidizing agents, which are listed below. The midpoint potential of a chemical measures how much energy is released when it is oxidized or reduced, with reducing agents having negative potentials and oxidizing agents positive potentials.

Respiratory enzymes and substrates in *E. coli*.[]

Respiratory enzyme	Redox pair	Midpoint potential (Volts)
Formate dehydrogenase	Bicarbonate / Formate	–0.43
Hydrogenase	Proton / Hydrogen	–0.42
NADH dehydrogenase	NAD^+ / NADH	–0.32
Glycerol-3-phosphate dehydrogenase	DHAP / Gly-3-P	–0.19
Pyruvate oxidase	Acetate + Carbon dioxide / Pyruvate	?
Lactate dehydrogenase	Pyruvate / Lactate	–0.19
D-amino acid dehydrogenase	2-oxoacid + ammonia / D-amino acid	?
Glucose dehydrogenase	Gluconate / Glucose	–0.14
Succinate dehydrogenase	Fumarate / Succinate	+0.03
Ubiquinol oxidase	Oxygen / Water	+0.82
Nitrate reductase	Nitrate / Nitrite	+0.42
Nitrite reductase	Nitrite / Ammonia	+0.36
Dimethyl sulfoxide reductase	DMSO / DMS	+0.16
Trimethylamine *N*-oxide reductase	TMAO / TMA	+0.13
Fumarate reductase	Fumarate / Succinate	+0.03

As shown above, *E. coli* can grow with reducing agents such as formate, hydrogen, or lactate as electron donors, and nitrate, DMSO, or oxygen as acceptors.[] The larger the difference in midpoint potential between an oxidizing and reducing agent, the more energy is released when they react. Out of these compounds, the succinate/fumarate pair is unusual, as its midpoint potential is close to zero. Succinate can therefore be oxidized to fumarate if a strong oxidizing agent such as oxygen is available, or fumarate can be reduced to succinate using a strong reducing agent such as formate. These alternative reactions are catalyzed by succinate dehydrogenase and fumarate reductase, respectively.[49]

Some prokaryotes use redox pairs that have only a small difference in midpoint potential. For example, nitrifying bacteria such as *Nitrobacter* oxidize nitrite to nitrate, donating the electrons to oxygen. The small amount of energy released in this reaction is enough to pump protons and generate ATP, but not enough to produce NADH or NADPH directly for use in anabolism.[50] This problem is solved by using a nitrite oxidoreductase to produce enough proton-motive force to run part of the electron transport chain in reverse, causing complex I to generate NADH.[51] [52]

Prokaryotes control their use of these electron donors and acceptors by varying which enzymes are produced, in response to environmental conditions.[53] This flexibility is possible because different oxidases and reductases use the same ubiquinone pool. This allows many combinations of enzymes to function together, linked by the common ubiquinol intermediate.[] These respiratory chains therefore have a modular design, with easily interchangeable sets of enzyme systems.

In addition to this metabolic diversity, prokaryotes also possess a range of isozymes – different enzymes that catalyze the same reaction. For example, in *E. coli*, there are two different types of ubiquinol oxidase using oxygen as an electron acceptor. Under highly aerobic conditions, the cell uses an oxidase with a low affinity for oxygen that can transport two protons per electron. However, if levels of oxygen fall, they switch to an oxidase that transfers only one proton per electron, but has a high affinity for oxygen.[54]

ATP synthase (complex V)

ATP synthase, also called *complex V*, is the final enzyme in the oxidative phosphorylation pathway. This enzyme is found in all forms of life and functions in the same way in both prokaryotes and eukaryotes.[] The enzyme uses the energy stored in a proton gradient across a membrane to drive the synthesis of ATP from ADP and phosphate (P_i). Estimates of the number of protons required to synthesize one ATP have ranged from three to four,[55] [56] with some suggesting cells can vary this ratio, to suit different conditions.[57]

This phosphorylation reaction is an equilibrium, which can be shifted by altering the proton-motive force. In the absence of a proton-motive force, the ATP synthase reaction will run from right to left, hydrolyzing ATP and pumping protons out of the matrix across the membrane. However, when the proton-motive force is high, the reaction is forced to run in the opposite direction; it proceeds from left to right, allowing protons to flow down their concentration gradient and turning ADP into ATP.[] Indeed, in the closely related vacuolar type H+-ATPases, the hydrolysis reaction is used to acidify cellular compartments, by pumping protons and hydrolysing ATP.[58]

ATP synthase is a massive protein complex with a mushroom-like shape. The mammalian enzyme complex contains 16 subunits and has a mass of approximately 600 kilodaltons.[59] The portion embedded within the membrane is called F_O and contains a ring of c subunits and the proton channel. The stalk and the ball-shaped headpiece is called F_1 and is the site of ATP synthesis. The ball-shaped complex at the end of the F_1 portion contains six proteins of two different kinds (three α subunits and three β subunits), whereas the "stalk" consists of one protein: the γ subunit, with the tip of the stalk extending into the ball of α and β subunits.[60] Both the α and β subunits bind nucleotides, but only the β subunits catalyze the ATP synthesis reaction. Reaching along the side of the F_1 portion and back into the membrane is a long rod-like subunit that anchors the α and β subunits into the base of the enzyme.

As protons cross the membrane through the channel in the base of ATP synthase, the F_O proton-driven motor rotates.[61] Rotation might be caused by changes in the ionization of amino acids in the ring of c subunits causing electrostatic interactions that propel the ring of c subunits past the proton channel.[62] This rotating ring in turn drives the rotation of the central axle (the γ subunit stalk) within the α and β subunits. The α and β subunits are prevented from rotating themselves by the side-arm, which acts as a stator. This movement of the tip of the γ subunit within the ball of α and β subunits provides the energy for the active sites in the β subunits to undergo a cycle of movements that produces and then releases ATP.[]

This ATP synthesis reaction is called the *binding change mechanism* and involves the active site of a β subunit cycling between three states.[] In the "open" state, ADP and phosphate enter the active site (shown in brown in the diagram). The protein then closes up around the molecules and binds them loosely – the "loose" state (shown in red). The enzyme then changes shape again and forces these molecules together, with the active site in the resulting "tight" state (shown in pink) binding the newly produced ATP molecule with very high affinity. Finally, the active site cycles back to the open state, releasing ATP and binding more ADP and phosphate, ready for the next cycle.

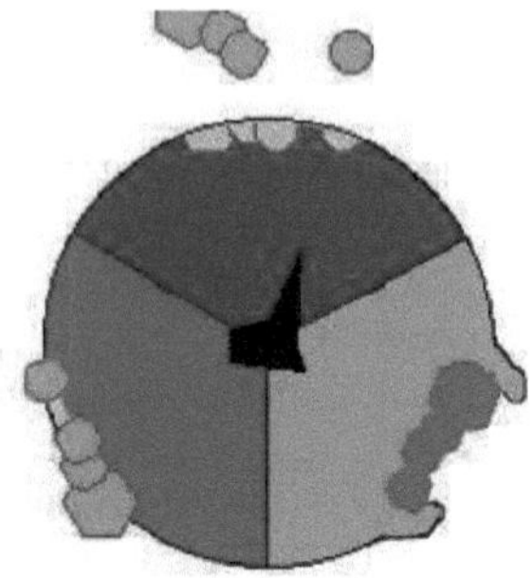

Mechanism of ATP synthase. ATP is shown in red, ADP and phosphate in pink and the rotating γ subunit in black.

In some bacteria and archaea, ATP synthesis is driven by the movement of sodium ions through the cell membrane, rather than the movement of protons.[63] [] Archaea such as *Methanococcus* also contain the A_1A_o synthase, a form of the enzyme that contains additional proteins with little similarity in sequence to other bacterial and eukaryotic ATP synthase subunits. It is possible that, in some species, the A_1A_o form of the enzyme is a specialized sodium-driven ATP synthase,[64] but this might not be true in all cases.[]

Reactive oxygen species

Molecular oxygen is an ideal terminal electron acceptor because it is a strong oxidizing agent. The reduction of oxygen does involve potentially harmful intermediates.[] Although the transfer of four electrons and four protons reduces oxygen to water, which is harmless, transfer of one or two electrons produces superoxide or peroxide anions, which are dangerously reactive.

These reactive oxygen species and their reaction products, such as the hydroxyl radical, are very harmful to cells, as they oxidize proteins and cause mutations in DNA. This cellular damage might contribute to disease and is proposed as one cause of aging.[65] [66]

The cytochrome c oxidase complex is highly efficient at reducing oxygen to water, and it releases very few partly reduced intermediates; however small amounts of superoxide anion and peroxide are produced by the electron transport chain.[] Particularly important is the reduction of coenzyme Q in complex III, as a highly reactive ubisemiquinone free radical is formed as an intermediate in the Q cycle. This unstable species can lead to electron "leakage" when electrons transfer directly to oxygen, forming superoxide.[67] As the production of reactive oxygen species by these proton-pumping complexes is greatest at high membrane potentials, it has been proposed that mitochondria regulate their activity to maintain the membrane potential within a narrow range that balances ATP production against oxidant generation.[68] For instance, oxidants can activate uncoupling proteins that reduce membrane potential.[69]

To counteract these reactive oxygen species, cells contain numerous antioxidant systems, including antioxidant vitamins such as vitamin C and vitamin E, and antioxidant enzymes such as superoxide dismutase, catalase, and peroxidases,[] which detoxify the reactive species, limiting damage to the cell.

Inhibitors

There are several well-known drugs and toxins that inhibit oxidative phosphorylation. Although any one of these toxins inhibits only one enzyme in the electron transport chain, inhibition of any step in this process will halt the rest of the process. For example, if oligomycin inhibits ATP synthase, protons cannot pass back into the mitochondrion.[] As a result, the proton pumps are unable to operate, as the gradient becomes too strong for them to overcome. NADH is then no longer oxidized and the citric acid cycle ceases to operate because the concentration of NAD^+ falls below the concentration that these enzymes can use.

Compounds	Use	Effect on oxidative phosphorylation
Cyanide Carbon monoxide Azide	Poisons	Inhibit the electron transport chain by binding more strongly than oxygen to the Fe–Cu center in cytochrome c oxidase, preventing the reduction of oxygen.[70]
Oligomycin	Antibiotic	Inhibits ATP synthase by blocking the flow of protons through the F_o subunit.[]
CCCP 2,4-Dinitrophenol	Poisons	Ionophores that disrupt the proton gradient by carrying protons across a membrane. This ionophore uncouples proton pumping from ATP synthesis because it carries protons across the inner mitochondrial membrane.[71]
Rotenone	Pesticide	Prevents the transfer of electrons from complex I to ubiquinone by blocking the ubiquinone-binding site.[72]
Malonate and oxaloacetate		Competitive inhibitors of succinate dehydrogenase (complex II).[73]

Not all inhibitors of oxidative phosphorylation are toxins. In brown adipose tissue, regulated proton channels called uncoupling proteins can uncouple respiration from ATP synthesis.[74] This rapid respiration produces heat, and is particularly important as a way of maintaining body temperature for hibernating animals, although these proteins may also have a more general function in cells' responses to stress.[75]

History

The field of oxidative phosphorylation began with the report in 1906 by Arthur Harden of a vital role for phosphate in cellular fermentation, but initially only sugar phosphates were known to be involved.[76] However, in the early 1940s, the link between the oxidation of sugars and the generation of ATP was firmly established by Herman Kalckar,[77] confirming the central role of ATP in energy transfer that had been proposed by Fritz Albert Lipmann in 1941.[78] Later, in 1949, Morris Friedkin and Albert L. Lehninger proved that the coenzyme NADH linked metabolic pathways such as the citric acid cycle and the synthesis of ATP.[79]

For another twenty years, the mechanism by which ATP is generated remained mysterious, with scientists searching for an elusive "high-energy intermediate" that would link oxidation and phosphorylation reactions.[80] This puzzle was solved by Peter D. Mitchell with the publication of the chemiosmotic theory in 1961.[81] At first, this proposal was highly controversial, but it was slowly accepted and Mitchell was awarded a Nobel prize in 1978.[82] [83] Subsequent research concentrated on purifying and characterizing the enzymes involved, with major contributions being made by David E. Green on the complexes of the electron-transport chain, as well as Efraim Racker on the ATP synthase.[84] A critical step towards solving the mechanism of the ATP synthase was provided by Paul D. Boyer, by his development in 1973 of the "binding change" mechanism, followed by his radical proposal of rotational catalysis in 1982.[] [85] More recent work has included structural studies on the enzymes involved in oxidative phosphorylation by John E. Walker, with Walker and Boyer being awarded a Nobel Prize in 1997.[86]

See also

- Respirometry
- TIM/TOM Complex

References

[11] Medical CHEMISTRY Compendium. By Anders Overgaard Pedersen and Henning Nielsen. Aarhus University. 2008
[12] Efremov R.G., Baradaran R., & Sazanov L.A., (2010) The arcdhitecture of respiratory complex I, Nature 465, 441-445

Further reading

IntroductoryAdvanced

External links

General resources

- Animated diagrams illustrating oxidative phosphorylation (http://www.wiley.com/legacy/college/boyer/0470003790/animations/electron_transport/electron_transport.htm) Wiley and Co *Concepts in Biochemistry*
- On-line biophysics lectures (http://www.life.uiuc.edu/crofts/bioph354/) Antony Crofts, University of Illinois at Urbana-Champaign
- ATP Synthase (http://www.youtube.com/watch?v=PjdPTY1wHdQ) Graham Johnson

Structural resources

- PDB molecule of the month:
  - ATP synthase (http://www.rcsb.org/pdb/static.do?p=education_discussion/molecule_of_the_month/pdb72_1.html)
  - Cytochrome c (http://www.rcsb.org/pdb/static.do?p=education_discussion/molecule_of_the_month/pdb36_1.html)
  - Cytochrome c oxidase (http://www.rcsb.org/pdb/static.do?p=education_discussion/molecule_of_the_month/pdb5_1.html)
- Interactive molecular models at Universidade Fernando Pessoa:
  - NADH dehydrogenase (http://www2.ufp.pt/~pedros/anim/2frame-ien.htm)
  - succinate dehydrogenase (http://www2.ufp.pt/~pedros/anim/2frame-iien.htm)
  - Coenzyme Q - cytochrome c reductase (http://www2.ufp.pt/~pedros/anim/2frame-iiien.htm)
  - cytochrome c oxidase (http://www2.ufp.pt/~pedros/anim/2frame-iven.htm)

Circadian_clock

A **circadian clock**, or **circadian oscillator**, is a biochemical mechanism that oscillates with a period of 24 hours and is coordinated with the day-night cycle. Circadian clocks are the central mechanisms which drive circadian rhythms. They consist of three major components:

1. A central oscillator with a period of about 24 hours that keeps time
2. A series of input pathways to this central oscillator to allow entrainment of the clock
3. A series of output pathways tied to distinct phases of the oscillator that regulate overt rhythms in biochemistry, physiology, and behavior throughout an organism

The clock is reset as the environment changes through an organism's ability to sense external time cues of which the primary one is light. Circadian oscillators are ubiquitous in tissues of the body where they are synchronized by both endogenous and external signals to regulate transcriptional activity throughout the day in a tissue-specific manner.[] The circadian clock is interwined with most cellular metabolic processes and it is affected by organism aging.[1] The basic molecular mechanisms of the biological clock have been defined in vertebrate species, *Drosophila melanogaster*, plants, fungi, and bacteria.[] []

Recent studies employing genomic approaches have further elucidated understanding of the circadian oscillator mechanism by providing a large-scale view into an organism's network and genetic architecture.[] []

Transcriptional and translational control

Evidence for a genetic basis of circadian rhythms in higher eukaryotes began with the discovery of the *period* (*per*) locus in *Drosophila melanogaster* from forward genetic screens completed by Ron Konopka and Seymour Benzer in 1971.[] Through the analysis of *per* circadian mutants and additional mutations on *Drosophila* clock genes, it was demonstrated that there is an underlying generative molecular mechanism of the circadian clock that consists of a set of core clock genes and their protein products, which together participate in positive and negative autoregulatory feedback loops of transcription and translation. Core circadian clock genes are defined as genes whose protein products are necessary components for the generation and regulation of circadian rhythms. Similar mechanisms have been demonstrated in mammals and other organisms.[] []

Mammalian clocks

Several mammalian clock genes have been identified and characterized through experiments on animals harboring naturally occurring, chemically induced, and targeted knockout mutations, and various comparative genomic approaches. The majority of identified clock components are transcriptional activators or repressors that modulate protein stability and nuclear translocation, and create two interlocking feedback loops.[] In the primary feedback loop, members of the basic helix-loop-helix (bHLH)-PAS (Period-Arnt-Single-minded) transcription factor family, CLOCK and BMAL1, heterodimerize in the cytoplasm to form a complex that, following translocation to the nucleus, initiates transcription of target genes such as the core clock genes *Period* genes (*PER1*, *PER2*, and *PER3*) and two *Cryptochrome* genes (*CRY1* and *CRY2*). Negative feedback is achieved by PER:CRY heterodimers that translocate back to the nucleus to repress their own transcription by inhibiting the activity of the CLOCK:BMAL1 complexes.[] Another regulatory loop is induced when CLOCK:BMAL1 heterodimers activate the transcription of Rev-ErbA and Rora, two retinoic acid-related orphan nuclear receptors. REV-ERBa and RORa subsequently compete to bind retinoic acid-related orphan receptor response elements (ROREs) present in Bmal1 promoter. Through the subsequent binding of ROREs, members of ROR and REV-ERB are able to regulate *Bmal1*. While RORs activate transcription of *Bmal1*, REV-ERBs repress the same transcription process. Hence, the circadian oscillation of *Bmal1* is both positively and negatively regulated by RORs and REV-ERBs.[]

Other organisms

In *D. melanogaster*, the gene cycle (CYC) is the orthologue of BMAL1 in mammals. Thus, CLOCK–CYC dimers activate the transcription of circadian genes. The gene timeless (TIM) is the orthologue for mammalian CRYs as the inhibitor; *D. melanogaster* CRY functions as a photoreceptor instead. In flies, CLK–CYC binds to the promoters of circadian-regulated genes only at the time of transcription. A stabilizing loop also exists where the gene vrille (VRI) inhibits whereas PAR-domain protein-1 (PDP1) activates Clock transcription.[] In *N. crassa*, the clock mechanism is analogous, but non-orthologous, to that of mammals and flies.[]

Post-translational modification

The autoregulatory feedback loops in clocks take about 24 hour to complete a cycle and constitute a circadian molecular clock. This generation of the ~24-hour molecular clock is governed by post-translational modifications such as phosphorylation, sumolyation, histone acetylation and methylation, and ubiquitination.[] Reversible phosphorylation regulates important processes such as nuclear entry, formation of protein complexes and protein degradation. Each of these processes significantly contributes to the delay that keeps the period at ~24 hours and lends the precision of a circadian clock by affecting the stability of aforementioned core clock proteins. Thus, while transcriptional regulation generates rhythmic RNA levels, regulated posttranslational modifications control protein abundance, subcellular localization, and repressor activity of PER and CRY.[]

Proteins responsible for post-translational modification of clock genes include casein kinase I family members (CSNK1D and CSNK1E) and the F-box and leucine-rich repeat protein 3 (FBXL3).[] In mammals, Casein kinase 1 epsilon and Casein kinase 1 delta are critical factors that regulate the core circadian protein turnover.[] Experimental manipulation on either of these proteins results in dramatic effects on circadian periods, such as altered kinase activities and cause shorter circadian periods, and further demonstrates the importance of the post-translational regulation within the core mechanism of the circadian clock.[] These mutations have become of particular interest in humans as they are implicated in familial advanced sleep phase syndrome.[] A small ubiquitin-related modifier protein modification of BMAL1 has also been proposed as another level of post-translational regulation.[]

Regulation of circadian oscillators

Circadian oscillators are simply oscillators with a period of approximately 24 hours. In response to light stimulus the body corresponds with a system and network of pathways that work together to determine the biological day and night. The regulatory networks involved in keeping the clock precise span over a range of post-translation regulation mechanisms. Circadian oscillators may be regulated by phosphorylation, SUMOylation, ubiquitination, and histone acetylation and deacetylation, the covalent modification of the histone tail which controls the level of chromatin structures causing the gene to be expressed more readily. Methylation of a protein structure adds a methyl group and regulates the protein function or gene expression and in histone methylation gene expression is either suppressed or activated through changing the DNA sequence. Histones go through an acetylation, methylation and phosphorylation process but the major structural and chemical changes are when enzymes *histone acetyltransferases* (HAT) and *histone deacetylases* (HDAC) add or remove acetyl groups from the histone causing a major change in DNA expression. By changing DNA expression, histone acetylation and methylation regulate how the circadian oscillator operates.

Systems biology approaches to elucidate oscillating mechanisms

Modern experimental approaches using systems biology have identified many novel components in biological clocks that suggest an integrative view on how organisms maintain circadian oscillation.[] []

Recently, Baggs et al. developed a novel strategy termed "Gene Dosage Network Analysis" (GDNA) to describe network features in the human circadian clock that contribute to an organism's robustness against genetic perturbations.[] In their study, the authors used small interfering RNA (siRNA) to induce dose-dependent changes in gene expression of clock components within immortalized human osteosarcoma U2OS cells in order to build gene association networks consistent with known biochemical constraints in the mammalian circadian clock. Employing multiple doses of siRNA powered their quantitative PCR to uncover several network features of the circadian clock, including proportional responses of gene expression, signal propagation through interacting modules, and compensation through gene expression changes.

Proportional responses in downstream gene expression following siRNA-induced perturbation revealed levels of expression that were actively altered with respect to the gene being knocked down. For example, when Bmal1 was knocked down in a dose-dependent manner, Rev-ErbA alpha and Rev-ErbA beta mRNA levels were shown to decrease in a linear, proportional manner. This supported previous findings that Bmal directly activates Rev-erb genes and further suggests Bmal as a strong contributor to Rev-erb expression.

In addition, the GDNA method provided a framework to study biological relay mechanisms in circadian networks through which modules communicate changes in gene expression.[] The authors observed signal propagation through interactions between activators and repressors, and uncovered unidirectional paralog compensation among several clock gene repressors—for example, when PER1 is depleted, there is an increase in Rev-erbs, which in turn propagates a signal to decrease expression in BMAL1, the target of the Rev-erb repressors.

By examining knockdown of several transcriptional repressors, GDNA also revealed paralog compensation where gene paralogs were upregulated through an active mechanism by which gene function is replaced following knockdown in a nonredunant manner—that is, one component is sufficient to sustain function. These results further suggested that a clock network utilizes active compensatory mechanisms rather than simple redundancy to confer robustness and maintain function. In essence, the authors proposed that the observed network features act in concert as a genetic buffering system to maintain clock function in the face of genetic and environmental perturbation.[] Following this logic, we may use genomics to explore network features in the circadian oscillator.

Another study conducted by Zhang et al. also employed a genome-wide small interfering RNA screen in U2OS cell line to identify additional clock genes and modifiers using luciferase reporter gene expression.[] Knockdown of nearly 1000 genes reduced rhythm amplitude. The authors found and confirmed hundreds of potent effects on period length or increased amplitude in secondary screens. Characterization of a subset of these genes demonstrated a dosage-dependent effect on oscillator function. Protein interaction network analysis showed that dozens of gene products directly or indirectly associate with known clock components. Pathway analysis revealed these genes are overrepresented for components of insulin and hedgehog signaling pathway, the cell cycle, and folate metabolism. Coupled with data demonstrating that many of these pathways are clock-regulated, Zhang et al. postulated that the clock is interconnected with many aspects of cellular function.

A systems biology approach may relate circadian rhythms to cellular phenomena that were not originally considered regulators of circadian oscillation.

Notes and references

External links

- Circadian screen search (http://biogps.gnf.org/circadian/)

Nitrogen_assimilation

Nitrogen assimilation is the formation of organic nitrogen compounds like amino acids from inorganic nitrogen compounds present in the environment. Organisms like plants, fungi and certain bacteria that cannot fix nitrogen gas (N_2) depend on the ability to assimilate nitrate or ammonia for their needs. Other organisms, like animals, depend entirely on organic nitrogen from their food.

Nitrogen assimilation in plants

Plants absorb nitrogen from the soil in the form of nitrate (NO_3^-) and ammonia (NH_3). In aerobic soils where nitrification can occur, nitrate is usually the predominant form of available nitrogen that is absorbed.[1] [] However this need not always be the case as ammonia can predominate in grasslands[2] and in flooded, anaerobic soils like rice paddies.[] Plant roots themselves can affect the abundance of various forms of nitrogen by changing the pH and secreting organic compounds or oxygen.[] This influences microbial activities like the inter-conversion of various nitrogen species, the release of ammonia from organic matter in the soil and the fixation of nitrogen by non-nodule-forming bacteria.

Ammonium ions are absorbed by the plant via ammonia transporters. Nitrate is taken up by several nitrate transporters that use a proton gradient to power the transport.[3] [4] Nitrogen is transported from the root to the shoot via the xylem in the form of nitrate, dissolved ammonia and amino acids. Usually[5] (but not always[6]) most of the nitrate reduction is carried out in the shoots while the roots reduce only a small fraction of the absorbed nitrate to ammonia. Ammonia (both absorbed and synthesized) is incorporated into amino acids via the glutamine synthetase-glutamate synthase (GS-GOGAT) pathway.[7] While nearly all[] the ammonia in the root is usually incorporated into amino acids at the root itself, plants may transport significant amounts of ammonium ions in the xylem to be fixed in the shoots.[8] This may help avoid the transport of organic compounds down to the roots just to carry the nitrogen back as amino acids.

Nitrate reduction is carried out in two steps. Nitrate is first reduced to nitrite (NO_2^-) in the cytosol by nitrate reductase using NADH or NADPH.[4] Nitrite is then reduced to ammonia in the chloroplasts (plastids in roots) by a ferredoxin dependent nitrite reductase. In photosynthesizing tissues, it uses an isoform of ferredoxin (Fd1) that is reduced by PSI while in the root it uses a form of ferredoxin (Fd3) that has a less negetive midpoint potential and can be reduced easily by NADPH.[9] In non photosynthesizing tissues, NADPH is generated by glycolysis and the pentose phosphate pathway.

In the chloroplasts,[10] glutamine synthetase incorporates this ammonia as the amide group of glutamine using glutamate as a substrate. Glutamate synthase (Fd-GOGAT and NADH-GOGAT) transfer the amide group onto an 2-oxoglutarate molecule producing two glutamates. Further transaminations are carried out make other amino acids (most commonly aspargine) from glutamine. While the enzyme glutamate dehydrogenase (GDH) does not play a direct role in the assimilation, it protects the mitochondrial functions during periods of high nitrogen metabolism and takes part in nitrogen remobilization.[11]

pH and Ionic balance during nitrogen assimilation

Every nitrate ion reduced to ammonia produces one OH^- ion. To maintain a pH balance, the plant must either excrete it into the surrounding medium or neutralize it with organic acids. This results in the medium around the plants roots becoming alkaline when they take up nitrate.

To maintain ionic balance, every NO_3^- taken into the root must be accompanied by either the uptake of a cation or the excretion of an anion. Plants like tomatoes take up metal ions like K^+, Na^+, Ca^{2+} and Mg^{2+} to exactly match every nitrate taken up and store these as the salts of organic acids like malate and oxalate.[12] Other plants like the soyabean balance most of their NO_3^- intake with the excretion of OH^- or HCO_3^-.[13]

Plants that reduce nitrates in the shoots and excrete alkali from their roots need to transport the alkali in an inert form from the roots to the shoots. To achieve this they synthesize malic acid in the leaves from neutral precursors like carbohydrates. The potassium ions brought to the leaves along with the nitrate in the xylem are then sent along with the malate to the roots via the phloem. In the roots, the malate is consumed. When malate is converted back to malic acid prior to use, an OH^- is released and excreted. ($RCOO^- + H_2O \rightarrow RCOOH + OH^-$) The potassium ions are then recirculated up the xylem with fresh nitrate. Thus the plants avoid having to absorb and store excess salts and also transport the OH^-.[14]

Plants like castor reduce a lot of nitrate in the root itself, and excrete the resulting base. Some of the base produced in the shoots is transported to the roots as salts of organic acids while a small amount of the carboxylates are just stored in the shoot itself.[15]

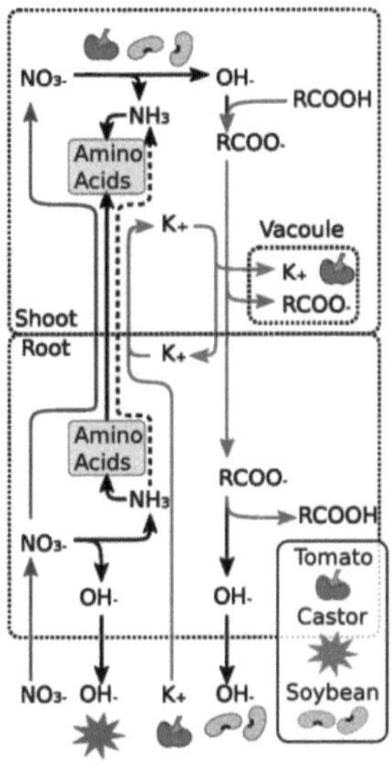

Different plants use different pathways to different levels. Tomatoes take in a lot of K^+ and accumulate salts in their vacuoles, castor reduces nitrate in the roots to a large extent and excretes the resulting alkali. Soy bean plants moves a large amount of malate to the roots where they convert it to alkali while the potassium is recirculated.

References

[1] Error: Bad DOI specified: 10.1146/annurev-arplant-042811-105532
[2] Error: Bad DOI specified: 10.1016/0038-0717(89)90152-1
[3] Error: Bad DOI specified: 10.1111/j.1365-3040.2011.02311.x
[4] Error: Bad DOI specified: 10.1046/j.1365-3040.2000.00595.x
[5] Error: Bad DOI specified: 10.1093/jxb/erf008
[6] Error: Bad DOI specified: 10.1111/j.1469-8137.1986.tb02905.x
[7] Error: Bad DOI specified: 10.1104/pp.105.071910
[8] Error: Bad DOI specified: 10.1093/jexbot/53.370.883
[9] Error: Bad DOI specified: 10.1104/pp.103.032755
[10] Error: Bad DOI specified: 10.1093/jxb/erm115
[11] Error: Bad DOI specified: 10.1016/S0981-9428(03)00060-3

Stress_(biology)

Stress is a person's response to a stressor such as an environmental condition or a stimulus. Stress is a body's way to react to a challenge. According to the stressful event, the body's way to respond to stress is by sympathetic nervous system activation which results in the fight-or-flight response. Stress typically describes a negative condition or a positive condition that can have an impact on a person's mental and physical well-being.

Stress and homeostasis

The term *stress* had none of its contemporary connotations before the 1920s. It is a form of the Middle English *destresse*, derived via Old French from the Latin *stringere*, "to draw tight."[1] The word had long been in use in physics to refer to the internal distribution of a force exerted on a material body, resulting in strain. In the 1920s and 1930s biological and psychological circles occasionally used the term to refer to a mental strain or to a harmful environmental agent that could cause illness. Walter Cannon used it in 1926 to refer to external factors that disrupted what he called homeostasis.[2] But "[...] Stress as an explanation of lived experience is absent from both lay and expert life narratives before the 1930s".[3]

Homeostasis is a concept central to the idea of stress. In biology, most biochemical processes strive to maintain equilibrium, a steady state that exists more as an ideal and less as an achievable condition. Environmental factors, internal or external stimuli, continually disrupt homeostasis; an organism's present condition is a state in constant flux moving about a homeostatic point that is that organism's optimal condition for living. Factors causing an organism's condition to diverge too far from homeostasis can be experienced as stress. A life-threatening situation such as a physical insult or prolonged starvation can greatly disrupt homeostasis. On the other hand, an organism's effortful attempt at restoring conditions back to or near homeostasis, often consuming energy and natural resources, can also be interpreted as stress. In such instances, an organism's fight-or-flight response recruits the body's energy stores and focuses attention to overcome the challenge at hand.

The ambiguity in defining this phenomenon was first recognized by Hans Selye (1907-1982) in 1926. In 1951 a commentator loosely summarized Selye's view of stress as something that "...in addition to being itself, was also the cause of itself, and the result of itself."[4] [5] First to use the term in a biological context, Selye continued to define stress as "the non-specific response of the body to any demand placed upon it". As of 2011 neuroscientists such as Bruce McEwen and Jaap Koolhaas believe that stress, based on years of empirical research, "should be restricted to conditions where an environmental demand exceeds the natural regulatory capacity of an organism".[6] Despite the numerous definitions given to stress, homeostasis appears to lie at its core.

Biological background

Stress can have profound effects on human biological systems.[7] Biology primarily attempts to explain major concepts of stress using a stimulus-response paradigm, broadly comparable to how a psychobiological sensory system operates. The central nervous system (brain and spinal cord) plays a crucial role in the body's stress-related mechanisms. Whether one should interpret these mechanisms as the body's response to a stressor or embody the act of stress itself is part of the ambiguity in defining what exactly stress is. Nevertheless, the central nervous system works closely with the body's endocrine system to regulate these mechanisms. One branch of the central nervous system, the sympathetic nervous system, becomes primarily active during a stress response, regulating many of the body's physiological functions in ways that ought to make an organism more adaptive to its environment. Below there follows a brief biological background of neuroanatomy and neurochemistry and how they relate to stress.

Neuroanatomy

Brain

The brain plays a critical role in the body's perception of and response to stress. However, pinpointing exactly which regions of the brain are responsible for particular aspects of a stress response is difficult and often unclear. Understanding that the brain works in more of a network-like fashion carrying information about a stressful situation across regions of the brain (from cortical sensory areas to more basal structures and vice versa) can help explain how stress and its negative consequences are heavily rooted in neural communication dysfunction. In spite of this, several important brain structures implicated in playing key roles in stress response pathways are described below.

Hypothalamus

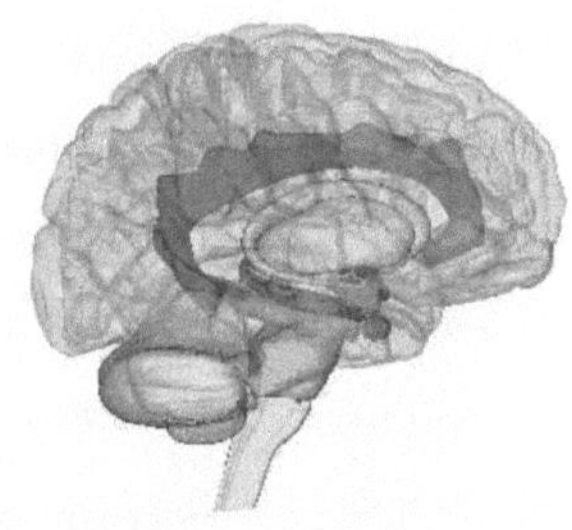

Human brain (hypothalamus=red, amygdala=green, hippocampus/fornix=blue, pons=gold, pituitary gland=pink)

The hypothalamus is a small portion of the brain located below the thalamus and above the brainstem. One of its most important functions is to help link together the body's nervous and endocrine systems. This structure has many bidirectional neural inputs and outputs from and to various other brain regions. These connections help regulate the hypothalamus' ability to secrete hormones into the body's blood stream, having far-reaching and long-lasting effects on physiological processes such as metabolism. During a stress response, the hypothalamus secretes various hormones, namely corticotropin-releasing hormone, which stimulates the body's pituitary gland and initiates a heavily regulated stress response pathway.

Amygdala

The amygdala is a small, "almond"-shaped structure located bilaterally, deep within the medial temporal lobes of the brain and is a part of the brain's limbic system, with projections to and from the hypothalamus, hippocampus, and locus coeruleus, among other areas. Thought to play a role in the processing of emotions, the amygdala has been implicated in modulating stress response mechanisms, particularly when feelings of anxiety or fear are involved.

Hippocampus

The hippocampus is a structure located bilaterally, deep within the medial temporal lobes of the brain, just lateral to each amygdala, and is a part of the brain's limbic system. The hippocampus is thought to play an important role in memory formation. There are numerous connections to the hippocampus from the cerebral cortex, hypothalamus, and amygdala, among other regions. During stress, the hippocampus is particularly important, in that cognitive processes such as prior memories can have a great influence on enhancing, suppressing, or even independently generating a stress response. The hippocampus is also an area in the brain that is susceptible to damage brought upon by chronic stress.

Locus coeruleus

The locus coeruleus is an area located in the pons of the brainstem that is the principal site of the synthesis of the neurotransmitter norepinephrine, which plays an important role in the sympathetic nervous system's fight-or-flight response to stress. This area receives input from the hypothalamus, amygdala, and raphe nucleus among other regions and projects widely across the brain as well as to the spinal cord.

Raphe nucleus

The raphe nucleus is an area located in the pons of the brainstem that is the principal site of the synthesis of the neurotransmitter serotonin, which plays an important role in mood regulation, particularly when stress is associated with depression and anxiety. Projections extend from this region to widespread areas across the brain, namely the hypothalamus, and are thought to modulate an organism's circadian rhythm and sensation of pain among other processes.

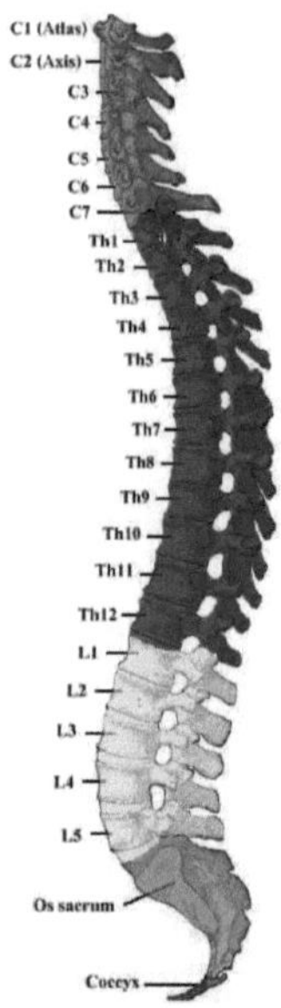

Human spinal cord

Spinal cord

The spinal cord plays a critical role in transferring stress response neural impulses from the brain to the rest of the body. In addition to the neuroendocrine blood hormone signaling system initiated by the hypothalamus, the spinal cord communicates with the rest of the body by innervating the peripheral nervous system. Certain nerves that belong to the sympathetic branch of the central nervous system exit the spinal cord and stimulate peripheral nerves, which in turn engage the body's major organs and muscles in a fight-or-flight manner.

Pituitary gland

The pituitary gland is a small organ that is located at the base of the brain just under the hypothalamus. This gland releases various hormones that play significant roles in regulating homeostasis. During a stress response, the pituitary gland releases hormones into the blood stream, namely adrenocorticotropic hormone, which modulates a heavily regulated stress response system

Adrenal gland

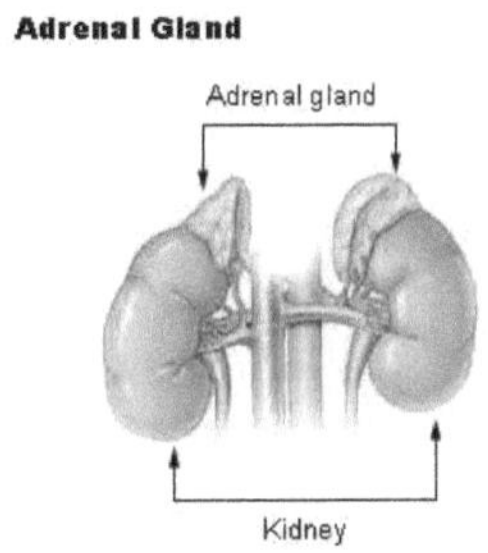

Adrenal gland

The adrenal gland is a major organ of the endocrine system that is located directly on top of the kidneys and is chiefly responsible for the synthesis of stress hormones that are released into the blood stream during a stress response. Cortisol is the major stress hormone released by the adrenal gland.

In addition to the locus coeruleus existing as a source of the neurotransmitter norepinephrine within the central nervous system, the adrenal gland can also release norepinephrine during a stress response into the body's blood stream, at which point norepinephrine acts as a hormone in the endocrine system.

Neurochemistry

Corticotropin-releasing hormone

Corticotropin-releasing hormone is the neurohormone secreted by the hypothalamus during a stress response that stimulates the anterior lobe of the pituitary gland by binding to its corticotropin-releasing hormone-receptors, causing the anterior pituitary to release adrenocorticotropic hormone.

Adrenocorticotropic hormone

Adrenocorticotropic hormone is the hormone secreted by the anterior lobe of the pituitary gland into the body's blood stream that stimulates the cortex of the adrenal gland by binding to its adrenocorticotropic hormone-receptors, causing the adrenal gland to release cortisol.

Cortisol

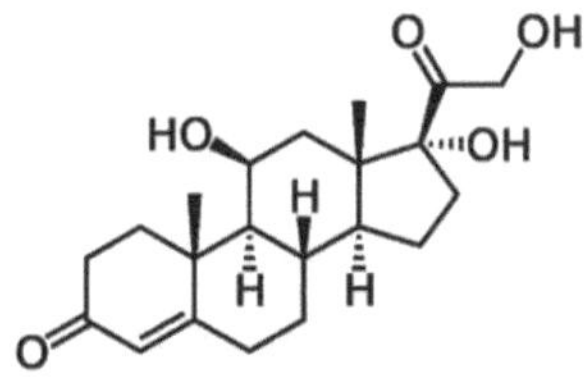
Cortisol

Cortisol is a steroid hormone, belonging to a broader class of steroids called glucocorticoids, produced by the adrenal gland and secreted during a stress response. Its primary function is to redistribute energy (glucose) to regions of the body that need it most (i.e., the brain and major muscles during a fight-or-flight situation). As a part of the body's fight-or-flight response, cortisol also acts to suppress the body's immune system.

Cortisol is synthesized from cholesterol in the adrenal cortex.[8] Its primary function is to increase blood sugar through gluconeogenesis, suppress the immune system and aid in fat and protein metabolism.[9]

Norepinephrine

Norepinephrine is a neurotransmitter released from locus coeruleus when stimulated by the hypothalamus during a stress response. Norepinephrine serves as the primary chemical messenger of the central nervous system's sympathetic branch that prepares the body for fight-or-flight response.

Serotonin

Serotonin is a neurotransmitter synthesized in the raphe nucleus of the pons of the brainstem and projects to most brain areas. Serotonin is thought to play an important role in mood regulation. Stress-induced serotonin dysfunctions have been associated with anxiety, fear, and depression-like symptoms.

Neuropeptide Y

Neuropeptide Y is a protein that is synthesized in the hypothalamus and acts as a chemical messenger in the brain. Traditionally, it has been thought to play an important role in appetite, feeding behavior, and satiety, but more recent findings have implicated Neuropeptide Y in anxiety and stress, specifically, stress resiliency.[10]

Biological mechanisms

Hypothalamic-pituitary-adrenal (HPA) axis

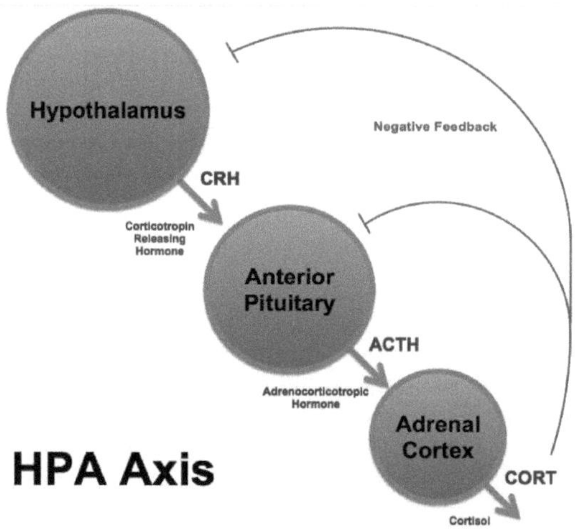

Basic hypothalamic–pituitary–adrenal axis summary (corticotropin-releasing hormone=CRH, adrenocorticotropic hormone=ACTH).

The HPA axis is a multi-step biochemical pathway where information is transmitted from one area of the body to the next via chemical messengers. Each step in this pathway, as in many biochemical pathways, not only passes information along to stimulate the next region but also receives feedback from messengers produced later in the pathway to either enhance or suppress earlier steps in the pathway – this is one way a biochemical pathway can regulate itself, via a feedback mechanism.

When the hypothalamus receives signals from one of its many inputs (e.g., cerebral cortex, limbic system, visceral organs) about conditions that deviate from an ideal homeostatic state (e.g., alarming sensory stimulus, emotionally charged event, energy deficiency), this can be interpreted as the initiation step of the stress-response cascade. The hypothalamus is stimulated by its inputs and then proceeds to secrete corticotropin-releasing hormones. This hormone is transported to its target, the pituitary gland, via the hypophyseal portal system (short blood vessels system), to which it binds and causes the pituitary gland to, in turn, secrete its own messenger, adrenocorticotropic hormone, systemically into the body's blood stream. When adrenocorticotropic hormone reaches and binds to its target, the adrenal gland, the adrenal gland in turn releases the final key messenger in the cascade, cortisol. Cortisol, once released, has widespread effects in the body. During an alarming situation in which a threat is detected and signaled to the hypothalamus from primary sensory and limbic structures, cortisol is one way the brain instructs the body to attempt to regain homeostasis – by redistributing energy (glucose) to areas of the body that need it most, that is, toward critical organs (the heart, the brain) and away from digestive and reproductive organs, during a potentially harmful situation in an attempt to overcome the challenge at hand.

After enough cortisol has been secreted to best restore homeostasis and the body's stressor is no longer present or the threat is no longer perceived, the heightened levels of cortisol in the body's blood stream eventually circulate to the pituitary gland and hypothalamus to which cortisol can bind and inhibit, essentially turning off the HPA-axis' stress-response cascade via feedback inhibition. This prevents additional cortisol from being released. This is biologically identified as a normal, healthy stress mechanism in response to a situation or stressor – a biological coping mechanism for a threat to homeostasis.

It is when the body's HPA-axis cannot overcome a challenge and/or is chronically exposed to a threat that this system becomes overtaxed and can be harmful to the body and brain. A second major effect of cortisol is to suppress the body's immune system during a stressful situation, again, for the purpose of redistributing metabolic resources primarily to fight-or-flight organs. While not a major risk to the body if only for a short period of time, if under chronic stress, the body becomes exceptionally vulnerable to immune system attacks. This is a biologically negative consequence of an exposure to a severe stressor and can be interpreted as stress in and of itself – a detrimental inability of biological mechanisms to effectively adapt to changes in homeostasis.

Immune response

Cortisol can weaken the activity of the immune system. Cortisol prevents proliferation of T-cells by rendering the interleukin-2 producer T-cells unresponsive to interleukin-1 (IL-1), and unable to produce the T-cell growth factor.[35]Cortisol also has a negative-feedback effect on interleukin-1.[36] IL-1 must be especially useful in combating some diseases; however, endotoxic bacteria have gained an advantage by forcing the hypothalamus to increase cortisol levels (forcing the secretion of CRH hormone, thus antagonizing IL-1). The suppressor cells are not affected by glucosteroid response-modifying factor (GRMF),[37] so the effective setpoint for the immune cells may be even higher than the setpoint for physiological processes (reflecting leukocyte redistribution to lymph nodes, bone marrow, and skin). Rapid administration of corticosterone (the endogenous Type I and Type II receptor agonist) orRU28362 (a specific Type II receptor agonist) to adrenalectomized animals induced changes in leukocytedistribution. Natural killer cells are not affected by cortisol.[38]

Effect of stress on the immune system

Stress is the body's reaction to any stimuli that disturb its equilibrium. When the equilibrium of various hormones is altered the effect of these changes can be detrimental to the immune system.[11] Much research has shown a negative effect stress has on the immune system, mostly through studies where participants were subjected to a variety of viruses. In one study, individuals caring for a spouse with dementia, representing the stress group, saw a significant decrease in immune response when given an influenza-virus vaccine compared to a non-stressed control group.[11][12] A similar study was conducted using a respiratory virus. Participants were infected with the virus and given a stress index. Results showed that an increase in score on the stress index correlated with greater severity of cold symptoms.[11] Studies with HIV have also shown stress to speed up viral progression. Men with HIV were 2–3 times more likely to develop AIDS when under above average stress.[11]

Effects of chronic stress

Chronic stress is defined as a "state of prolonged tension from internal or external stressors, which may cause various physical manifestations – e.g., asthma, back pain, arrhythmias, fatigue, headaches, HTN, irritable bowel syndrome, ulcers, and suppress the immune system". Chronic stress takes a more significant toll on your body than acute stress does. It can raise blood pressure, increase the risk of heart attack and stroke, increase vulnerability to anxiety and depression, contribute to infertility, and hasten the aging process. For example, results of one study demonstrated that individuals who reported relationship conflict lasting one month or longer have a greater risk of developing illness and show slower wound healing. Similarly, the effects that acute stressors have on the immune system may be increased when there is perceived stress and/or anxiety due to other events. For example, students who are taking exams show weaker immune responses if they also report stress due to daily hassles.[13]

Mechanisms of chronic stress

Studies revealing the relationship between the immune system and the central nervous system indicate that stress can alter the function of white blood cells involved in immune function, known as lymphocytes and macrophages. People undergoing stressful life events, such as marital turmoil or bereavement, have a weaker lymphoproliferative response. After antigens initiate an immune response, these white blood cells send signals, composed of cytokines and other hormonal proteins, to the brain and neuroendocrine system.[14] Cytokines are molecules involved with cell signaling. Cortisol, a hormone released during stressful situations, affects the immune system greatly by preventing the production of cytokines. During chronic stress, cortisol is over produced, causing fewer receptors to be produced on immune cells so that inflammation cannot be ended. A study involving cancer patient's parents confirmed this finding. Blood samples were taken from the participants. Researchers treated the samples of the parents of cancer patients with a cortisol-like substance and stimulated cytokine production. Cancer patient parents' blood was significantly less effective at stopping cytokine from being produced.[15]

Chronic stress and wound healing

The immune system also plays a role in stress and the early stages of wound healing. It is responsible for preparing tissue for repair and promoting recruitment of certain cells to the wound area.[13] Consistent with the fact that stress alters the production of cytokines, Graham et al. found that chronic stress associated with care giving for a person with Alzheimer's Disease leads to delayed wound healing. Results indicated that biopsy wounds healed 25% more slowly in the chronically stressed group, or those caring for a person with Alzheimer's disease.[11]

Chronic stress and development

Chronic stress has also been shown to impair developmental growth in children by lowering the pituitary gland's production of growth hormone, as in children associated with a home environment involving serious marital discord, alcoholism, or child abuse.[16]

Chronic stress and memory

Chronic stress is seen to affect parts of the brain where memories are processed through and stored. When people feel stressed, stress hormones get over-secreted, which affects the brain. This secretion is made up of glucocorticoids, including cortisol, which are steroid hormones that the adrenal gland releases.[17]

Stress and visceral fat

Studies of female monkeys at Wake Forest University (2009) discovered that individuals suffering from higher stress have higher levels of visceral fat in their bodies. This suggests a possible cause-and-effect link between the two, wherein stress promotes the accumulation of visceral fat, which in turn causes hormonal and metabolic changes that contribute to heart disease and other health problems.[]

Psychological concepts

Eustress

Selye published in 1975 a model dividing stress into eustress and distress.[18] Where stress enhances function (physical or mental, such as through strength training or challenging work), it may be considered eustress. Persistent stress that is not resolved through coping or adaptation, deemed distress, may lead to anxiety or withdrawal (depression) behavior.

The difference between experiences that result in *eustress* and those that result in *distress* is determined by the disparity between an experience (real or imagined) and personal expectations, and resources to cope with the stress. Alarming experiences, either real or imagined, can trigger a stress response.[19]

Coping

Responses to stress include adaptation, psychological coping such as stress management, anxiety, and depression. Over the long term, distress can lead to diminished health and/or increased propensity to illness; to avoid this, stress must be managed.

Stress management encompasses techniques intended to equip a person with effective coping mechanisms for dealing with psychological stress, with stress defined as a person's physiological response to an internal or external stimulus that triggers the fight-or-flight response. Stress management is effective when a person uses strategies to cope with or alter stressful situations.

There are several ways of coping with stress,[20] such as controlling the source of stress or learning to set limits and to say "no" to some of the demands that bosses or family members may make.

A person's capacity to tolerate the source of stress may be increased by thinking about another topic such as a hobby, listening to music, or spending time in a wilderness.

Cognitive appraisal

Lazarus[21] argued that, in order for a psychosocial situation to be stressful, it must be appraised as such. He argued that cognitive processes of appraisal are central in determining whether a situation is potentially threatening, constitutes a harm/loss or a challenge, or is benign.

Both personal and environmental factors influence this primary appraisal, which then triggers the selection of coping processes. Problem-focused coping is directed at managing the problem, whereas emotion-focused coping processes are directed at managing the negative emotions. Secondary appraisal refers to the evaluation of the resources available to cope with the problem, and may alter the primary appraisal.

In other words, primary appraisal includes the perception of how stressful the problem is and the secondary appraisal of estimating whether one has more than or less than adequate resources to deal with the problem that affects the overall appraisal of stressfulness. Further, coping is flexible in that, in general, the individual examines the effectiveness of the coping on the situation; if it is not having the desired effect, s/he will, in general, try different strategies.[]

Clinical symptoms and disorders

Symptoms

Signs of stress may be cognitive, emotional, physical, or behavioral.

Cognitive symptoms

- Memory problems
- Inability to concentrate
- Poor judgment
- Pessimistic approach or thoughts
- Anxious or racing thoughts
- Constant worrying

Emotional symptoms

- Moodiness
- Irritability or short temper
- Agitation, inability to relax
- Feeling overwhelmed
- Sense of loneliness and isolation
- Depression or general unhappiness

Physical symptoms

- Aches and pains
- Diarrhea or constipation
- Increased frequency of urination
- Indigestion
- Changes in blood glucose
- Nausea, dizziness
- Chest pain, rapid heartbeat
- Loss of sex drive
- Frequent colds
- Irregular periods.

Behavioral symptoms

- Eating more or less
- Sleeping too much or too little
- Isolating oneself from others
- Procrastinating or neglecting responsibilities
- Using alcohol, cigarettes, or drugs to relax
- Nervous habits (e.g. nail biting, pacing)

DSM-IV TR

Diagnosis

A renewed interest in salivary alpha amylase as a marker for stress has surfaced. Yamaguchi M, Yoshida H (2005) have analyzed a newly introduced hand-held device called the Cocorometer developed by Nipro Corp., Japan. They state that this can be reliably used to analyze the amylase levels and is definitely a cheaper alternative as compared to the more expensive ELISA kits. The working consists of a meter and a saliva collecting chip, which can be inserted into the meter to give the readings. The levels of amylase obtained have been calibrated according to standard population, and can be categorized into four levels of severity.[22]

Measuring stress level independent of differences in people's personalities has been inherently difficult: Some people are able to process many stressors simultaneously, while others can barely address a few. Such tests as the Trier Social Stress Test attempted to isolate the effects of personalities on ability to handle stress in a laboratory environment. Other psychologists, however, proposed measuring stress indirectly, through self-tests.

Because the amount of stressors in a person's life often (although not always) correlates with the amount of stress that person experiences, researchers combine the results of stress and burnout self-tests. Stress tests help determine the number of stressors in a person's life, while burnout tests determine the degree to which the person is close to the state of burnout. Combining both helps researchers gauge how likely additional stressors will make him or her experience mental exhaustion.[23]

Health risk factors

Both negative and positive stressors can lead to stress. The intensity and duration of stress changes depending on the circumstances and emotional condition of the person suffering from it (Arnold. E and Boggs. K. 2007). Some common categories and examples of stressors include:

- Sensory input such as pain, bright light, noise, temperatures, or environmental issues such as a lack of control over environmental circumstances, such as food, air and/or water quality, housing, health, freedom, or mobility.
- Social issues can also cause stress, such as struggles with conspecific or difficult individuals and social defeat, or relationship conflict, deception, or break ups, and major events such as birth and deaths, marriage, and divorce.
- Life experiences such as poverty, unemployment, clinical depression, obsessive compulsive disorder, heavy drinking,[] or insufficient sleep can also cause stress. Students and workers may face performance pressure stress from exams and project deadlines.
- Adverse experiences during development (e.g. prenatal exposure to maternal stress,[24] [25] poor attachment histories,[] sexual abuse)[26] are thought to contribute to deficits in the maturity of an individual's stress response systems. One evaluation of the different stresses in people's lives is the Holmes and Rahe stress scale.

Generalized anxiety syndrome

The areas of the brain affected by generalised anxiety disorder

During passive activity, patients with generalised anxiety disorder (GAD) exhibit increased metabolic rates in the occipital, temporal and frontal lobes and in the cerebellum and thalamus compared with healthy controls. Increased metabolic activity in the basal ganglia has also been reported in patients with GAD during vigilance tasks. These finding suggest that there may be hyperactive brain circuits in GAD.[27]

The areas of the brain affected in generalised anxiety disorder (advanced)

Patients with generalised anxiety disorder (GAD) exhibit increased metabolic rates in several brain regions compared with healthy controls. Hyperactive neurotransmitter circuits between the cortex, thalamus, amygdala and hypothalamus have been implicated in the disorder. Hypofunction of serotonergic neurones arising from the dorsal raphe nucleus and GABAergic neurones that are widely distributed in the brain may result in a lack of inhibitory effect on the putative GAD pathway. Furthermore, overactivity of noradrenergic neurones arising from the locus coeruleus may produce excessive excitation in the brain areas implicated in GAD.[28]

The septohippocampal circuit

Based on early neuroanatomical observations and studies with psychoactive drugs, the septohippocampal circuit has been proposed as a model for anxiety disorders. The circuit that links the septum, amygdala, hippocampus and fornix is thought to process external stimuli and regulate the behavioural response through wider projections in the brain. Hyperstimulation of this putative 'behavioural inhibition' circuit, through dysfunctional noradrenergic and serotonergic neurotransmission, has been implicated in producing anxiety, and increased arousal and attention.[29]

The noradrenaline pathways in generalised anxiety disorder

In generalised anxiety disorder (GAD) there is increased noradrenaline transmission from both the locus coeruleus and the caudal raphe nuclei. The locus coeruleus-noradrenaline system is associated with anxiety and may mediate the autonomic symptoms associated with stress such as increased heart rate, dilated pupils, tremour and sweating.[30]

Serotonergic pathways showing the effects of generalised anxiety disorder

Serotonergic nuclei are found in the rostral and caudal raphe nuclei. Neurones ascend from the rostral raphe nuclei to the cerebral cortex, limbic regions and basal ganglia. The activity of neurones innervating the pre-frontal cortex, basal ganglia and limbic region is decreased in generalised anxiety disorder (GAD). The activity of descending neurones from serotonergic nuclei in the brainstem is unaffected in GAD. This altered neurotransmitter balance contributes towards the feeling of anxiety associated with GAD.[31]

GABAergic pathways showing the effects of generalised anxiety disorder

GABA is the main inhibitory neurotransmitter in the central nervous system (CNS). GABAergic inhibition is seen at all levels of the CNS, including the hypothalamus, hippocampus, cerebral cortex and cerebellar cortex. The activity of GABAergic neurones is decreased in generalised anxiety disorder.[32]

Panic disorder

The areas of the brain affected in panic disorder

There are a number of areas of the brain affected in panic disorder. Decreased serotonin activity in the amygdala and frontal cortex induces symptoms of anxiety, whereas decreased activity in the periaqueductal gray results in defensive behaviours and postural freezing. The locus coeruleus increases norepinephrine release mediating physiological and behavioural arousal, while the hypothalamus mediates the sympathetic nervous system.[33] [34] [35]

The areas of the brain affected in panic disorder (advanced)

Hyperactive neurotransmitter circuits between the cortex, thalamus, hippocampus, amygdala, hypothalamus and periaqueductal gray matter have been implicated in panic disorder. Hypofunction of serotonergic neurones arising from the rostral raphe nucleus may result in a lack of inhibitory effect on the putative panic pathways in the brain. While, overactivity of norepinephrine neurons arising from the locus coeruleus may produce excessive excitation in the regions implicated in panic disorder. Physiological symptoms of the panic response are medicated by the autonomic nervous system through connections with the locus coeruleus and hypothalamus.[33] [34] [35] [36] [37]

The serotonin pathways in panic disorder

The principal serotonin centres in the brain are the caudal and rostral raphe nuclei. Transmission of serotonin from the rostral raphe nuclei to the pre-aquaductal grey, amygdala, temporal lobe and limbic cortex is decreased in panic disorder compared with normal. Serotonin transmission to other target regions of the brain remain unchanged.[38]

The norepinephrine pathways in panic disorder

In panic disorder there is increased norepinephrine transmission from both the locus coeruleus and the caudal raphe nuclei. The locus coeruleus-norepinephrine system may have a significant role in processing fear-related stimuli or it may affect fear-related processing by stimulating other regions of the brain implicated in anxiety and fear behaviours i.e. amygdala, hippocampus, hypothalamus, cortex and spinal cord.[38]

General adaptation syndrome

Physiologists define stress as how the body reacts to a stressor, real or imagined, a stimulus that causes stress. Acute stressors affect an organism in the short term; chronic stressors over the longer term.

Alarm is the first stage, which is divided into two phases: the *shock* phase and the *antishock* phase.[39]

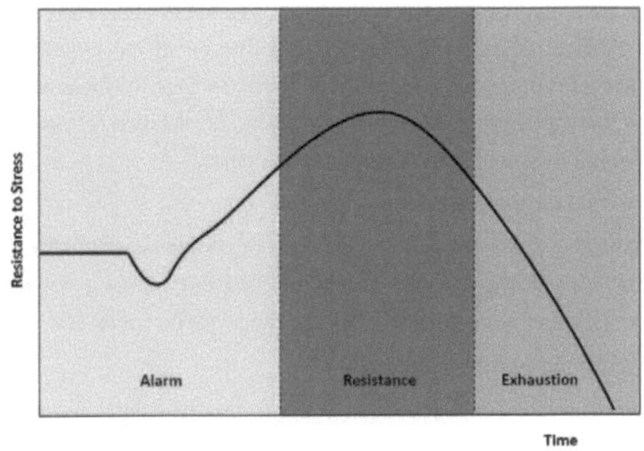

A diagram of the General Adaptation Syndrome model.

- *Shock phase*: During this phase, the body can endure changes such as hypovolemia, hypoosmolarity, hyponatremia, hypochloremia, hypoglycemia—the stressor effect. This phase resembles Addison's disease. The organism's resistance to the stressor drops temporarily below the normal range and some level of shock (e.g. circulatory shock) may be experienced.
- *Antishock phase*: When the threat or stressor is identified or realized, the body starts to respond and is in a state of alarm. During this stage, the locus coeruleus/sympathetic nervous system is activated and catecholamines such as adrenaline are being produced, hence the fight-or-flight response. The result is: increased muscular tonus, increased blood pressure due to peripheral vasoconstriction and tachycardia, and increased glucose in blood. There is also some activation of the HPA axis, producing glucocorticoids (cortisol, aka the S-hormone or stress-hormone).

Resistance is the second stage and increased secretion of glucocorticoids play a major role, intensifying the systemic response—they have lypolytic, catabolic and antianabolic effects: increased glucose, fat and aminoacid/protein concentration in blood. Moreover, they cause lymphocytopenia, eosinopenia, neutrophilia and polycythemia. In high doses, cortisol begins to act as a mineralocorticoid (aldosteron) and brings the body to a state similar to hyperaldosteronism. If the stressor persists, it becomes necessary to attempt some means of coping with the stress. Although the body begins to try to adapt to the strains or demands of the environment, the body cannot keep this up indefinitely, so its resources are gradually depleted.

The third stage could be either *exhaustion* or *recovery*:

- **Recovery** stage follows when the system's compensation mechanisms have successfully overcome the stressor effect (or have completely eliminated the factor which caused the stress). The high glucose, fat and aminoacid levels in blood prove useful for anabolic reactions, restoration of homeostasis and regeneration of cells.
- **Exhaustion** is the alternative third stage in the GAS model. At this point, all of the body's resources are eventually depleted and the body is unable to maintain normal function. The initial autonomic nervous system symptoms may reappear (sweating, raised heart rate, etc.). If stage three is extended, long-term damage may result (prolonged vasoconstriction results in ischemia which in turn leads to cell necrosis), as the body's immune system becomes exhausted, and bodily functions become impaired, resulting in decompensation.

The result can manifest itself in obvious illnesses, such as peptic ulcer and general trouble with the digestive system (e.g. occult bleeding, melena, constipation/obstipation), depression, diabetes, or even cardiovascular problems (angina pectoris), along with other mental illnesses.

Phobia

The areas of the brain affected in phobia

There are a number of areas of the brain affected in phobia. Activation of the amygdala causes anticipatory anxiety or avoidance (conditioned fear) while activation of the hypothalamus activates the sympathetic nervous system. Other regions of the brain involved in phobia include the thalamus and the cortical structures, which may form a key neural network along with the amygdala. Stimulation of the locus coeruleus increases noradrenaline release mediating physiological and behavioural arousal.[35]

The noradrenaline pathways in phobia

One hypothesis about the biological basis of phobia suggests that there is an excess of noradrenaline in the principal noradrenergic pathways in the brain and that this causes a down-regulation of post-synaptic adrenergic receptors. Transmission of noradrenaline from the caudal raphe nuclei and the locus coeruleus is increased in phobia.[40]

The serotonin pathways in phobia

The principal serotonin centres in the brain are the caudal and rostral raphe nuclei. Transmission of serotonin from the rostral raphe nuclei to the thalamus, limbic cortex and cerebral cortex is decreased in phobia compared with normal. The other major pathways for serotonin transmission which involve the basal ganglia and cerebellum, and project down the spinal cord, remain unchanged.[40]

Post-traumatic stress disorder (PTSD)

See article: Posttraumatic stress disorder

PTSD is a severe anxiety disorder that can develop after exposure to any event that results in psychological trauma. This event may involve the threat of death to oneself or to someone else, or to one's own or someone else's physical, sexual, or psychological integrity, overwhelming the individual's ability to cope. As an effect of psychological trauma, PTSD is less frequent and more enduring than the more commonly seen acute stress response. Diagnostic symptoms for PTSD include **intrusion, avoidance and hyperarousal** -- re-experiencing the original trauma(s) through "flashbacks" or nightmares (*intrusion*), emotional numbing or avoidance of stimuli associated with the trauma, and increased arousal, such as difficulty falling or staying asleep, anger, and hypervigilance. Formal diagnostic criteria (both DSM-IV-TR and ICD-10) require that the symptoms last more than one month and cause significant impairment in social, occupational, or other important areas of functioning.

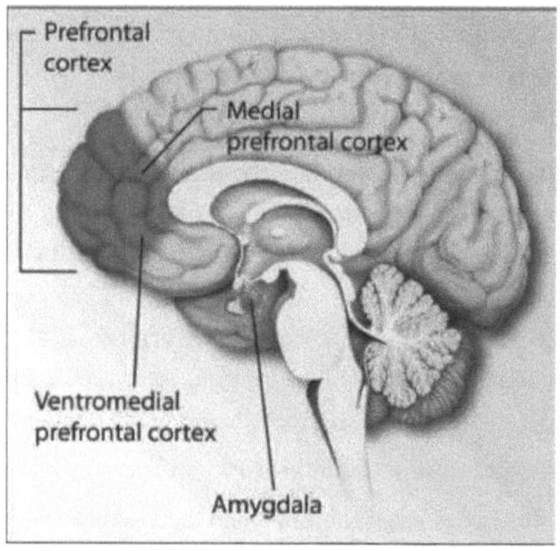

Regions of the brain associated with stress and posttraumatic stress disorder[41]

The areas of the brain affected in post-traumatic stress disorder

Sensory input, memory formation and stress response mechanisms are affected in patients with PTSD. The regions of the brain involved in memory processing that are implicated in PTSD include the hippocampus, amygdala and frontal cortex. While the heightened stress response is likely to involve the thalamus, hypothalamus and locus coeruleus.[36] [37]

Memory

Cortisol works with epinephrine (adrenaline) to create memories of short-term emotional events; this is the proposed mechanism for storage of flash bulb memories, and may originate as a means to remember what to avoid in the future. However, long-term exposure to cortisol damages cells in the hippocampus; this damage results in impaired learning. Furthermore, it has been shown that cortisol inhibits memory retrieval of already stored information.

Atrophy of the hippocampus in posttraumatic stress disorder

There is consistent evidence from MRI volumetric studies that hippocampal volume is reduced in posttraumatic stress disorder (PTSD). This atrophy of the hippocampus is thought to represent decreased neuronal density. However, other studies suggest that hippocampal changes are explained by whole brain atophy and generalised white matter atrophy is exhibited by people with PTSD.[42] [43]

Depression

The areas of the brain affected in depression

Many areas of the brain appear to be involved in depression including the frontal and temporal lobes and parts of the limbic system including the cingulate gyrus. However, it is not clear if the changes in these areas cause depression or if the disturbance occurs as a result of the etiology of psychiatric disorders.[44]

The hypothalamic-pituitary-adrenal (HPA) axis in depression

In depression, the hypothalamic-pituitary-adrenal (HPA) axis is upregulated with a down-regulation of its negative feedback controls. Corticotropin-releasing factor (CRF) is hypersecreted from the hypothalamus and induces the release of adrenocorticotropin hormone (ACTH) from the pituitary. ACTH interacts with receptors on adrenocortical cells and cortisol is released from the adrenal glands; adrenal hypertrophy can also occur. Release of cortisol into the circulation has a number of effects, including elevation of blood glucose. The negative feedback of cortisol to the hypothalamus, pituitary and immune system is impaired. This leads to continual activation of the HPA axis and excess cortisol release. Cortisol receptors become desensitized leading to increased activity of the pro-inflammatory immune mediators and disturbances in neurotransmitter transmission.[45] [46] [47] [48]

The serotonin pathways in depression

Serotonin transmission from both the caudal raphe nuclei and rostral raphe nuclei is reduced in patients with depression compared with non-depressed controls. Increasing the levels of serotonin in these pathways, by reducing serotonin reuptake and hence increasing serotonin function, is one of the therapeutic approaches to treating depression.[49]

The noradrenaline pathways in depression

In depression the transmission of noradrenaline is reduced from both of the principal noradrenergic centres – the locus coeruleus and the caudal raphe nuclei. An increase in noradrenaline in the frontal/prefrontal cortex modulates the action of selective noradrenaline reuptake inhibition and improves mood. Increasing noradrenaline transmission to other areas of the frontal cortex modulates attention.[50]

History in research

The current usage of the word *stress* arose out of Selye's 1930s experiments. He started to use the term to refer not just to the agent but to the state of the organism as it responded and adapted to the environment. His theories of a universal non-specific stress response attracted great interest and contention in academic physiology and he undertook extensive research programs and publication efforts.[51]

While the work attracted continued support from advocates of psychosomatic medicine, many in experimental physiology concluded that his concepts were too vague and unmeasurable. During the 1950s, Selye turned away from the laboratory to promote his concept through popular books and lecture tours. He wrote for both non-academic physicians and, in an international bestseller entitled *Stress of Life*, for the general public.

A broad biopsychosocial concept of stress and adaptation offered the promise of helping everyone achieve health and happiness by successfully responding to changing global challenges and the problems of modern civilization.

Selye coined the term "eustress" for positive stress, by contrast to distress. He argued that all people have a natural urge and need to work for their own benefit, a message that found favor with industrialists and governments.[51] He also coined the term *stressor* to refer to the causative event or stimulus, as opposed to the resulting state of stress.

From the late 1960s, academic psychologists started to adopt Selye's concept; they sought to quantify "life stress" by scoring "significant life events," and a large amount of research was undertaken to examine links between stress and disease of all kinds. By the late 1970s, stress had become the medical area of greatest concern to the general population, and more basic research was called for to better address the issue. There was also renewed laboratory research into the neuroendocrine, molecular, and immunological bases of stress, conceived as a useful heuristic not necessarily tied to Selye's original hypotheses. The US military became a key center of stress research, attempting to understand and reduce combat neurosis and psychiatric casualties.[51]

The psychiatric diagnosis *post-traumatic stress disorder* (*PTSD*) was coined in the mid-1970s, in part through the efforts of anti-Vietnam War activists and the Vietnam Veterans Against the War, and Chaim F. Shatan. The condition was added to the *Diagnostic and Statistical Manual of Mental Disorders* as *posttraumatic stress disorder* in 1980.[52] PTSD was considered a severe and ongoing emotional reaction to an extreme psychological trauma, and as such often associated with soldiers, police officers, and other emergency personnel. The stressor may involve threat to life (or viewing the actual death of someone else), serious physical injury, or threat to physical or psychological integrity. In some cases, it can also be from profound psychological and emotional trauma, apart from any actual physical harm or threat. Often, however, the two are combined.

By the 1990s, "stress" had become an integral part of modern scientific understanding in all areas of physiology and human functioning, and one of the great metaphors of Western life. Focus grew on stress in certain settings, such as workplace stress, and stress management techniques were developed. The term also became a euphemism, a way of referring to problems and eliciting sympathy without being explicitly confessional, just "stressed out." It came to cover a huge range of phenomena from mild irritation to the kind of severe problems that might result in a real breakdown of health. In popular usage, almost any event or situation between these extremes could be described as stressful.[1] [51]

See also

- Amygdala hijack
- Burnout (psychology)
- Chronic stress
- Compassion fatigue
- Defense Physiology
- Depression (differential diagnoses)
- Grief
- Psychological resilience
- Self-medication
- Stress (disambiguation)
- Stress management
- Trier Social Stress Test
- Work-life balance
- Xenohormesis

References

Notes

[1] Keil, R.M.K. (2004) Coping and stress: a conceptual analysis (http://www.ingentaconnect.com/content/bsc/jan/2004/00000045/00000006/art00013) *Journal of Advanced Nursing*, 45(6), 659–665

[2] W. B. Cannon. "Physiological regulation of normal states: some tentative postulates concerning biological homeostatics." IN: A. Pettit (ed.). A Charles Richet: ses amis, ses collègues, ses élèves, p. 91. Paris: Éditions Médicales, 1926.

[5] Compare *The Stress of Life*, Hans Selye, New York: McGraw-Hill, 1956.

[6] Koolhaas, J., et al. "Stress revisited: A critical evaluation of the stress concept." Neuroscience and Biobehavioral Reviews 35, 1291–1301, 2011.

[8] www.drexelmed.edu/documents/pharmacology_physiology/clinks/gpharmhandouts/Adrenal%2520Steroids%2520Grad%2520Pharm%25202004.doc+cortisol+s

[9] Measurement of Cortisol: Introduction to the hypothalamic - pituitary - adrenocortical axis. Nancy A. Nicolson, 2007

[10] Urban, Janice, et al. (2008). "Neuropeptide Y in the Amygdala Induces Long-Term Resilience to Stress-Induced Reductions in Social Responses But Not Hypothalamic–Adrenal–Pituitary Axis Activity or Hyperthermia" The Journal of Neuroscience, 28(4): 893–903; doi:10.1523/JNEUROSCI.0659-07.2008

[11] Khansari, D., Murgo, A., & Faith, R. (1990). Effects of stress on the immune system. Immunology Today, 11, 170–175.

[12] Kemeny, M. E. (2007). "Understanding the interaction between psychosocial stress and immune-related diseases: A stepwise progression." Brain, Behavior, and Immunity, 21 (8), 1009–1018.

[13] Graham, J., Christian, L. & Kiecolt-Glaser, J. (2006). Stress, Age, and Immune Function: Toward a Lifespan Approach. Journal of Behavioral Medicine, 29, 389–400.

[14] Dantzer, R. & Kelley, K. (1989). Stress and immunity: An integrated view of relationships between the brain and the immune system. Life Sciences, 44, 1995–2008.

[15] Glaser, R. & Kiecolt-Glaser, J. K. (2005). "Stress-induced immune dysfunction: Implications for health." Immunology, 5, 243–251.

[16] Powell, Brasel, & Blizzard, 1967.

[22] Shankar, A.A.; Dandekar, R.C. "Assessment of stress in patients with Recurrent Aphthous Stomatitis, by salivary alpha amylase using a Cocorometer" dissertation submitted for Oral Pathology to Maharashtra University of Health Sciences, Nashik in December 2009.

[23] Truby, William. "Stress Test" (http://www.trubyachievements.com/Business/Stress_Management/Stress_Test.html), *Stress Test: A self assessment*, December, 2009.

[24] Davis et al. (June 2007). Prenatal Exposure to Maternal Depression and Cortisol Influences Infant Temperament. *Journal of the American Academy of Child & Adolescent Psychiatry*, v46 n6 p737.

[25] O'Connor, Heron, Golding, Beveridge & Glover. (June 2002). Maternal antenatal anxiety and children's behavioural/emotional problems at 4 years. *Br J Psychiatry*. 180:478–9.

[26] Michael D. DeBellis, George P. Chrousos, Lorah D. Dorn, Lillian Burke, Karin Helmers, Mitchel A. Kling, Penelope K. Trickett, and Frank W. Putnam. Hypothalamic—Pituitary—Adrenal Axis Dysregulation in Sexually Abused Girls

[27] J Clin Psychiatry 2001;62 (Suppl. 11):22–27.

[28] J Clin Psychiatry 2001;62 (Suppl. 11):22–27

[29] L'Encéphale 1983;IX:161B–166B.

[30] Anxiolytics and sedative-hypnotics. In: Essential Psychopharmacology. Neuroscientific basis and practical applications. Stahl S. Cambridge, UK: Cambridge University Press, 1996: 167–215.

[31] Psychiatr Clin North Am 2001;21:75–97.

[32] J Clin Psychiatry 2001;62:22–27.

[33] J Clin Psychiatry 1997;58 (suppl. 2):4–11.

[34] J Clin Psychiatry 1998;59 (suppl. 8):24–28.

[35] Am J Psychiatry 2000;157:493–505.

[36] Current Opinions in Neurobiology 2000;10:211–218.

[37] J Clin Psychiatry 2000;61 (Suppl. 5):24–29.

[38] J Clin Psychiatry 1997;58 (Suppl. 2):4–11.

[39] Pathology: Medical student's library ISBN 978-83-61047-18-6 http://books.google.rs/books?id=PvkOMoLNjgYC&lpg=PA272&ots=PuqI645Scs&dq=antishock%20phase%20general%20adaptive%20syndrome&pg=PA272#v=onepage&q=antishock%20phase%20general%20adaptive%20syndrome&f=false

[40] Drug treatments for obsessive compulsive disorder, panic disorder, and phobia. In: Essential Psychopharmacology. Neuroscientific basis and practical applications. Stahl S. Cambridge, UK: Cambridge University Press, 1996: 216–248.

[42] Biol Psychiatry 2002;52:119–125.

[43] Semin Clin Neuropsychiatry 2001;6:113–145.

[44] Depression. What happens in the brain? www.brainexplorer.org Accessed 20th May 2002.

[45] Primary Care Companion J Clin Psychiatry 2001;3:151–155.

[46] Metabolism 2002;51:5–10.

[47] Neuroendocrinol 2001;13:1009–1023.

[48] Br Med Bull 1996;52:597–617.
[49] Neural plasticity and disorders of the nervous system. In: Neuroscience at a glance, Barker RA and Barasi S with Neal MJ. London, England, 1999:105–121.
[50] Neurotransmitters and psychiatric disorders. In: Psychiatric disorders with a biochemical basis. Donaldson D. Carnthorn, UK: The Parthenon Publishing Group Inc, 1998: 45–55.
[51] Viner, R. (1999) Putting Stress in Life: [[Hans Selye (http://www.jstor.org/pss/285410)] and the Making of Stress Theory]. *Social Studies of Science*, Vol. 29, No. 3 (June 1999), pp. 391–410
[52] ; on-line (http://www.istss.org/what/history2.cfm).

Bibliography

- Carlson, N.R. & Heth, C.D. (2007). *Psychology the science of behaviour*. 4th ed. Upper Saddle River, New Jersey: Pearson Education, Inc., 527.
- Petersen, C.; Maier, S.F.; Seligman, M.E.P. (1995). *Learned Helplessness: A Theory for the Age of Personal Control*. New York: Oxford University Press. ISBN 0-19-504467-3
- Seligman, M.E.P. (1975). *Helplessness: On Depression, Development, and Death*. San Francisco: W.H. Freeman. ISBN 0-7167-2328-X
- Seligman, M.E.P. (1990). *Learned Optimism*. New York: Knopf. (Reissue edition, 1998, Free Press, ISBN 0-671-01911-2).
- Holmes, T.H. and Rahe, R.H. (1967). The social readjustments rating scales. *Journal of Psychosomatic Research* 11:213–218.

External links

- The American Institute of Stress (http://www.stress.org/topic-definition-stress.htm)
- University of Massachusetts Medical School Stress Reduction Program (http://www.umassmed.edu/Content.aspx?id=41254)
- "Research on Work-Related Stress" (http://osha.europa.eu/publications/reports/203/index.htm), European Agency for Safety and Health at Work (EU-OSHA)
- "Working on Stress" (http://osha.europa.eu/publications/magazine/5/index.htm), European Agency for Safety and Health at Work (EU-OSHA)
- "Taming Stress" (http://www.sciam.com/article.cfm?chanID=sa006&articleID=00083A00-318C-1F30-9AD380A84189F2D7), *Scientific American*, September 2003
- Self help guide (http://www.nhsdirect.nhs.uk/help/bodykey/questions/index.aspx?nodes=u0NJCORqOyc=) (NHS Direct)

Plastid

The **plastid** (Greek: πλαστός; plastós: formed, molded – plural **plastids**) is a major organelle found in the cells of plants and algae. Plastids are the site of manufacture and storage of important chemical compounds used by the cell. They often contain pigments used in photosynthesis, and the types of pigments present can change or determine the cell's color. They possess a double-stranded DNA molecule, which is circular, like that of prokaryotes.

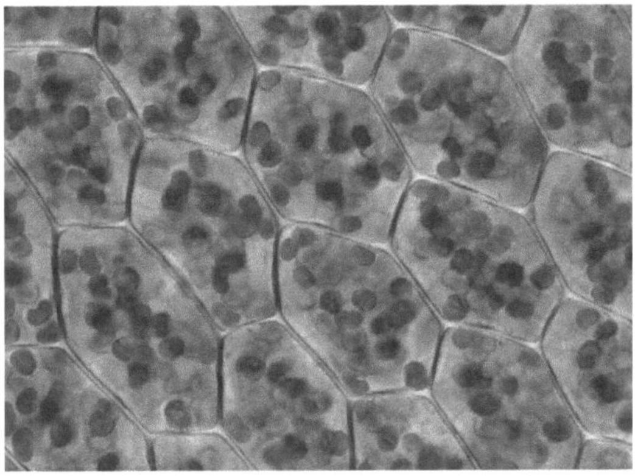

Plant cells with visible chloroplasts.

Plastids in plants

Plastids carry out photosynthesis, the storage of products like starch and the synthesis of many classes of molecules such as fatty acids and terpenes which are used for energy production and as raw material for the synthesis of other molecules. For example, the components of the plant cuticle and epicuticular waxes are synthesized in the epidermal cells from palmitic acid synthesized in the chloroplasts of mesophyll cells.[1] All plastids are derived from proplastids (formerly "eoplasts", *eo-*: dawn, early), which are present in the meristematic regions of the plant. Proplastids and young chloroplasts commonly divide by binary fission, but more mature chloroplasts also have this capacity.

In plants, plastids may differentiate into several forms, depending upon which function they play in the cell. Undifferentiated plastids (*proplastids*) may develop into any of the following variants:[]

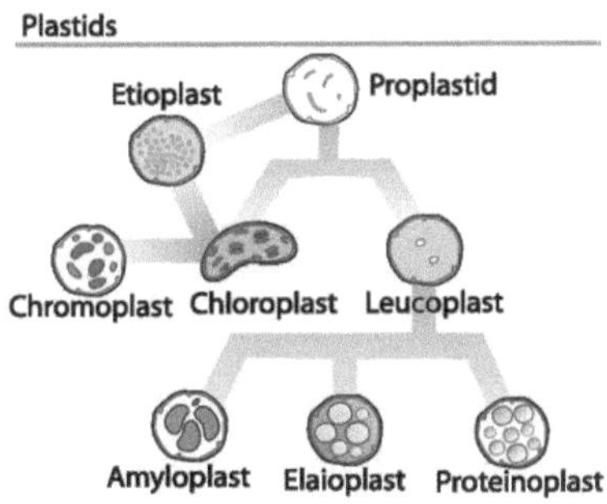

- Chloroplasts green plastids: for photosynthesis; *see also etioplasts, the predecessors of chloroplasts*
- Chromoplasts coloured plastids: for pigment synthesis and storage
- Gerontoplasts: control the dismantling of the photosynthetic apparatus during senescence
- Leucoplasts colourless plastids: for monoterpene synthesis; *leucoplasts sometimes differentiate into more specialized plastids:*
  - Amyloplasts: for starch storage and detecting gravity
  - Elaioplasts: for storing fat
  - Proteinoplast/aleuroplasts: for storing and modifying protein

Depending on their morphology and function, plastids have the ability to differentiate, or redifferentiate, between these and other forms.

Each plastid creates multiple copies of a circular 75–250 kilobase plastome. The number of genome copies per plastid is variable, ranging from more than 1000 in rapidly dividing cells, which, in general, contain few plastids, to 100 or fewer in mature cells, where plastid divisions have given rise to a large number of plastids. The plastome contains about 100 genes encoding ribosomal and transfer ribonucleic acids (rRNAs and tRNAs) as well as proteins involved in photosynthesis and plastid gene transcription and translation. However, these proteins only represent a small fraction of the total protein set-up necessary to build and maintain the structure and function of a particular type of plastid. Plant Nuclear genes encode the vast majority of plastid proteins, and the expression of plastid genes and nuclear genes is tightly co-regulated to coordinate proper development of plastids in relation to cell differentiation.

Plastid DNA exists as large protein-DNA complexes associated with the inner envelope membrane and called 'plastid nucleoids'. Each nucleoid particle may contain more than 10 copies of the plastid DNA. The proplastid contains a single nucleoid located in the centre of the plastid. The developing plastid has many nucleoids, localized at the periphery of the plastid, bound to the inner envelope membrane. During the development of proplastids to chloroplasts, and when plastids convert from one type to another, nucleoids change in morphology, size and location within the organelle. The remodelling of nucleoids is believed to occur by modifications to the composition and abundance of nucleoid proteins.

Many plastids, particularly those responsible for photosynthesis, possess numerous internal membrane layers.

In plant cells, long thin protuberances called stromules sometimes form and extend from the main plastid body into the cytosol and interconnect several plastids. Proteins, and presumably smaller molecules, can move within stromules. Most cultured cells that are relatively large compared to other plant cells have very long and abundant stromules that extend to the cell periphery.

Plastids in algae

In algae, the term leucoplast is used for all unpigmented plastids and their function differs from the leucoplasts of plants. Etioplasts, amyloplasts and chromoplasts are plant-specific and do not occur in algae. Plastids in algae and hornworts may also differ from plant plastids in that they contain pyrenoids.

Glaucocystophytic algae contain muroplasts, which are similar to chloroplasts except that they have a cell wall that is similar to that of prokaryotes. Rhydophytic algae contain rhydoplasts, which are red chloroplasts that allow the algae to photosynthesise to a depth of up to 268 m.[]

Inheritance of plastids

Most plants inherit the plastids from only one parent. In general, angiosperms inherit plastids from the female gamete, whereas many gymnosperms inherit plastids from the male pollen. Algae also inherit plastids from only one parent. The plastid DNA of the other parent is, thus, completely lost.

In normal intraspecific crossings (resulting in normal hybrids of one species), the inheritance of plastid DNA appears to be quite strictly 100% uniparental. In interspecific hybridisations, however, the inheritance of plastids appears to be more erratic. Although plastids inherit mainly maternally in interspecific hybridisations, there are many reports of hybrids of flowering plants that contain plastids of the father. Approximately 20% of angiosperms, including alfalfa (*Medicago sativa*), normally show biparental inheritance of plastids.[2]

Origin of plastids

Plastids are thought to have originated from endosymbiotic cyanobacteria. The symbiosis evolved around 1500 million years ago and enabled eukaryotes to carry out oxygenic photosynthesis.[3] Three evolutionary lineages have since emerged in which the plastids are named differently: chloroplasts in green algae and plants, rhodoplasts in red algae and muroplasts in the glaucophytes. The plastids differ by their pigmentation, but also in ultrastructure. The chloroplasts, e.g., have lost all phycobilisomes, the light harvesting complexes found in cyanobacteria, red algae and glaucophytes, but instead contain stroma and grana thylakoids, structures found only in plants and in closely related green algae. The glaucocystophycean plastid — in contrast to the chloroplasts and the rhodoplasts — is still surrounded by the remains of the cyanobacterial cell wall. All these primary plastids are surrounded by two membranes.

Complex plastids start by secondary endosymbiosis, when a eukaryote engulfs a red or green alga and retains the algal plastid, which is typically surrounded by more than two membranes. In some cases these plastids may be reduced in their metabolic and/or photosynthetic capacity. Algae with complex plastids derived by secondary endosymbiosis of a red alga include the heterokonts, haptophytes, cryptomonads, and most dinoflagellates (= rhodoplasts). Those that endosymbiosed a green alga include the euglenids and chlorarachniophytes (= chloroplasts). The Apicomplexa, a phylum of obligate parasitic protozoa including the causative agents of malaria (*Plasmodium* spp.), toxoplasmosis (*Toxoplasma gondii*), and many other human or animal diseases also harbor a complex plastid (although this organelle has been lost in some apicomplexans, such as *Cryptosporidium parvum*, which causes cryptosporidiosis). The 'apicoplast' is no longer capable of photosynthesis, but is an essential organelle, and a promising target for antiparasitic drug development.

Some dinoflagellates and sea slugs, in particular of the genus Elysia, take up algae as food and keep the plastid of the digested alga to profit from the photosynthesis; after a while, the plastids are also digested. These captured plastids are known as kleptoplastids.

Sources

- A Novel View of Chloroplast Structure [4]: contains fluorescence images of chloroplasts and stromules as well as an easy to read chapter.
- PDF [5]

See also

- Mitochondrion

References

[1] Kolattukudy, PE (1996) Biosynthetic pathways of cutin and waxes, and their sensitivity to environmental stresses. In: Plant Cuticles. Ed. by G. Kerstiens, BIOS Scientific publishers Ltd., Oxford, pp 83-108
[2] Error: Bad DOI specified: 10.1007/s10265-009-0291-z
[4] http://www.plantphys.net/article.php?ch=e&id=122
[5] http://www.hos.ufl.edu/ctdcweb/Birky01.pdf

Further reading

- Gould SB, Waller RR, McFadden GI (2008). Plastid evolution. Annu Rev Plant Biol 59: 491–517.
- Tree of Life Eukaryotes (http://tolweb.org/Eukaryotes/3)

External links

- Transplastomic plants for biocontainment (biological confinement of transgenes) (http://www.coextra.eu/projects/project199.html) — Co-extra research project on coexistence and traceability of GM and non-GM supply chains

Article Sources and Contributors

Synechocystis *Source*: http://en.wikipedia.org/w/index.php?title=Synechocystis *Contributors*: Adrian J. Hunter, Bilardi, Flakinho, Galootius, IceCreamAntisocial, Kleopatra, McFjodor, RE73, Rjwilmsi, Several Times, Verme005, 7 anonymous edits

Cyanobacteria *Source*: http://en.wikipedia.org/w/index.php?title=Cyanobacteria *Contributors*: *Drillzor*, 2D, ABF, AXRL, Abanima, Abhinav.pratap.iitd, AceHide, Activision45, AdamRetchless, Addshore, Aitias, Alansohn, Alice.haugen, Allens, Allisonh12, Alshon3, Andonic, Andrea105, Andres, Angryc, Anthere, Aperium, Apothecia, Arcadian, Arkuat, ArnoldReinhold, AxelBoldt, Azhyd, Barneca, Basa, BatteryIncluded, Ben Ben, Ben Webber, Bender235, Bendzh, Bgpaulus, Billinghurst, BlackAndy, Blizzard youkai, Bob Blaylock, Bobet, Bomac, Bowlhover, Br77rino, Bradcallaghan23, Brastein, Bryan Derksen, CWii, Cabus, Cacycle, Calabe1992, Caltas, CanadianLinuxUser, CanisRufus, Carl Hirschie Johnson, Celestechang, Cephal-odd, Chameleon, Chet Gray, ChrisHodgesUK, Claystu, Cohesion, Colonies Chris, CommonsDelinker, Cosmicharade, DO11.10, Dacoutts, Daggerstab, Dale Gerdemann, Dan Koehl, Daniel Vaulot, Danny Yee, Darktremor, Darrien, David Ardell, David Little, Daycd, DeadEyeArrow, Deflective, Delldot, DemosDemon, DerHexer, Deryck Chan, Dhanasooraj, Dimketr, Dirac1933, Dr.Bastedo, Draeco, Dragonlv8, Drbogdan, Drestros power, Drmatthew 123, Dsumm438, Dvdfwr, Dysprosia, E Wing, Eaefremov, EauOo, Ebyabe, Ekrub-ntyh, Ellywa, Empco, EnSamulili, EncycloPetey, Epastore, Epbr123, Epipelagic, Erianna, Eric Kvaalen, Estevezj, Eugene van der Pijll, Evano1van, Excirial, Fadix, Fatal exception, Favonian, Filovirus, Flewis, Fluck, Fluxbyte, Flyguy649, ForestDim, Franco3450, Fratrep, Freederick, Gadfium, Gareth Griffith-Jones, GeoGreg, Giftlite, Gikü, Ginsengbomb, Glenn, Global Birder, Gorank4, Greg Hedberg, Grumpy Fisherman, Guil sobrinho, HMSSolent, Haham hanuka, Hans Dunkelberg, Harizotoh9, Hedles, Hellbus, Henrygb, Hut 8.5, Huw Powell, HydrogenGuy, IanOsgood, Ibinthinkin, ImperfectlyInformed, Incnis Mrsi, Into vertigo, IrishJew, Isaacdealey, Isopropyl, J. Spencer, J.delanoy, JBurkhardt10, JCrue, JDP90, JSRD421, Jackhynes, Janke, Jebus989, Jeff schiller, JidGom, Jilkmarine, Jmv2009, Joe Decker, John, John Cumbers, Jon186, Josh Grosse, Karmosin, Katanada, Kathovo, Kbh3rd, Kelvinsong, Kevintoodlepoot, Kim grace, Kingdon, Kleopatra, Kokoriko, Kraftlos, Kupirijo, LOL, Lascorz, Legoktm, Lfstevens, Ling.Nut, Linnell, Little Mountain 5, Lottelw12, Lrunge, Lxtc, M1ss1ontomars2k4, MER-C, Mac, Mad Greg, Madhero88, Magnus Manske, MarcoTolo, Marshman, Materialscientist, Mathonius, Matthewjparker, Mbbachelor, Mccready, Mctowell, Mdjohnson, Menchi, Metron4, Mgiganteus1, Michaeldsuarez, Mifter, Mike Rosoft, Mintleaf, Mksoley, Mordekai wiki, Moshe Constantine Hassan Al-Silverburg, Msiami, Mysid, N2e, Nakon, Narayanese, NebuchadnezzarN, Nehrams2020, Nelsonwill, Netrapt, NewEnglandYankee, Niceguyedc, Nihiltres, NimbusWeb, Nolelover, Nutster, Onco p53, Onnager, Originalwana, Osborne, Ouedbirdwatcher, Palica, Panic, Leeroy, PatríciaR, Paul A, Pedroflecha, Perdita, Petrolmaps, Phaedriel, Pharaoh of the Wizards, Philip Trueman, Picaroon, Piki, Pinethicket, Pinky sl, Pion, Plantsurfer, Poeloq, PoeticVerse, Popotatatoto, Portillo, Postglock, Ppgardne, Praxis, Pseudofusulina, Quintote, RA0808, RW Marloe, Radon210, Rails, Rainbowwrasse, Raven in Orbit, Richard001, Rjwilmsi, Rkitko, Robert UU, Robinh, Rotational, Rowan Adams, Rsduhamel, Saigon punkid, Samtram, Sceptre, SchuminWeb, Scwlong, Sdittami, Seglea, Sellyme, Senachild, Shadowshador, Shaurya gaur, Shayno, Shniken, Sidhekin, Sigpro47, Silentlight, Sinn, Slon02, Smartse, Smith609, Smithbrenon, Smooth O, Sophus Bie, Spartan.mike, SpikeToronto, Squidonius, Squids and Chips, Ss6803, Stemonitis, StephenDawson, Sunray, Tabletop, TakuyaMurata, Tawker, Tbhotch, TestPilot, ThatPeskyCommoner, The Thing That Should Not Be, TheAlphaWolf, Theda, Tide rolls, TimBentley, Tipiac, Tiptoety, Tmasters18, Tommy2010, Touchstone42, Twinsday, Unyoyega, UrsoBR, Valich, Velella, Versus22, Warofdreams, Wavelength, Whoop whoop pull up, Wiki13, William Avery, Wilson44691, Wlodzimierz, Woohookitty, WriterHound, Yersinia, Zanze123, Zashaw, Zoe, Zorknot, Ásgeir IV., आशिष भटनागर, 532 anonymous edits

Heterotroph *Source*: http://en.wikipedia.org/w/index.php?title=Heterotroph *Contributors*: 2001:470:1F04:155E:0:0:0:2, 2help, AB, Aatomic1, Accurizer, Airplaneman, Aitias, Alan Liefting, Alloquep, Andres, Anomalocaris, Anthere, Anxietycello, Ascend, Astronaut, Berestede, Bklynsk8er92, Borgx, Brewhaha@edmc.net, Brya, Caltas, CanisRufus, Capricorn42, Cburnett, Cfsenel, CharlieEchoTango, ChrisGualtieri, Cinik, Closedmouth, Correogsk, Cssaunders, DSRH, DarkEvil, Darkwind, Davidiad, Dchall1, Deflective, Deli nk, Determinate, Donald Albury, Dr.K., Dsmurat, ELApro, Edwinstearns, El C, Eleassar, Emw, EncycloPetey, Epbr123, Epipelagic, Excirial, FinalRapture, Fluffernutter, FlyingToaster, Freshjivaz, Furuba9, Fvasconcellos, Ghazer, Gise-354x, Glenn, Gruzd, Gwil, Hans Dunkelberg, Helljumper43, Humanisticmystic, IRP, Id1337x, Illuminatiscott, Intelcore7, InverseHypercube, Isaac Oh, Jay2332, JerrySteal, Jesseblum, Jmeeter, Jmeppley, Jncraton, JoshuaZ, Jplasti, Julian Mendez, KnowledgeOfSelf, KnowledgeRequire, L Kensington, Lcarscad, Lexor, Logan, Luckas Blade, M2afilikeballs, Marek69, MarnetteD, Mattbr, Maxamiliangil, Mdjohnson, Michael Snow, Mikael Häggström, Mion, Mristroph, Mrt3366, Mrusso18, Mysdaao, NellieBly, Netrapt, Nyttend, Obli, OlEnglish, Onco p53, Opabinia regalis, OsamaBinLogin, Oscar, Paleorthid, Palosirkka, Paullb, Pekaje, Peregrine981, Pianoplonkers, Pickilomeeni, Plumbago, Plumpurple, Pmokeefe, Polinizador, Prolog, Quintote, Ragesoss, Rcknight, Redmarkviolinist, Richard001, Ringler, Rjwilmsi, Robodoc.at, Ronaldac, Roseindela, Saippuakauppias, Samw, Samwb123, Screensaver7, Sebsmoot, Senator Palpatine, Serouj, Shenme, Shlomke, Shrumster, Slavering.dog, Slon02, Snoito, Someone the Person, Spartan-James, Spongefrog, Svetovid, TheDude2006, Thejadefalcon, Theopolisme, Timo Honkasalo, Ukilledanoob, Versus22, Wavelength, Wickey-nl, Winterst, Wirtlo, WriterHound, Youssefsan, ماني, اشف عقيل, 248 anonymous edits

Photosynthesis *Source*: http://en.wikipedia.org/w/index.php?title=Photosynthesis *Contributors*: (jarbarf), *drew, -- April, 1exec1, 28421u2232nfenfcenc, 678910, 9258fahsflkh917fas, @pple, AJRM, ASKingquestions, ATMarsden, Abductive, Abu-Fool Danyal ibn Amir al-Makhiri, Academic Challenger, Accuruss, Ace of Spades, Acefox, Adashiel, Adenosine, Agathman, Ahoerstemeier, Aidentbrown, Aitias, Alan Liefting, Alansohn, Alexius08, Ali, Ali salter, Alksub, Amatulic, Amercer09, AmiDaniel, Amorymeltzer, Andris, AndyZ, Angam, Angela, Angr, AnonMoos, Anonymi, Ans, Antandrus, AnthonyQBachler, Antony2, Apokryltaros, Aranea Mortem, Arcadian, Arjun01, Art LaPella, At09kg, Atrian, Aut1221, Avicennasis, AxelBoldt, B820, BD2412, BGFMSM, Bailey654, BanyanTree, Barras, Bateman1234, Bayerischermann, Bbatsell, Beetstra, Ben-Zin, Bencherlite, Bendzh, Bento00, Bergsten, Bertiethecat, Bfahome, Bhadani, Billare, Blahed768, Bluedeed, Bobo192, Bodnotbod, Boghog, Boing! said Zebedee, Bonadea, Bongwarrior, Bookandcoffee, Bowlhover, Brainmachine, BrightStarSky, Bryan Derksen, Burnhamd, CERminator, CQJ, CWY2190, Cacycle, Calabe1992, Call me Bubba, Caltas, CalumH93, Camw, Can't sleep, clown will eat me, CanadianLinuxUser, Canterbury Tail, CardinalDan, Cassius987, CatPath, Causa sui, Cayte, Cbrown1023, Cdorman2, Celarnor, CerealKiller, Ceyockey, Chamal N, Chao, Chasingsol, Cheesy Yeast, Chepry, Chirpy7, Chlor, Chodorkovskiy, ChrisGualtieri, ChrisHodgesUK, Chrislk02, Chun-hian, Cimex, Ciphergoth, Ckatz, Clicketyclack, ClockworkSoul, ClockworkTroll, Closedmouth, Clubjuggle, Clyde10race, Cman12345657578, Cmdrjameson, Coching, Cohesion, Collegedegree, Color probe, ComCat, Commander, CommonsDelinker, Condem, Connormah, Consequencefree, Conversion script, Corrigen, Courcelles, Crazychinchilla, Crazytales, Creidieki, Crittens, Crystallina, CurtisSwain, Cyde, Cyrius, DARTH SIDIOUS 2, DVD R W, DaGizza, Dacrycarpus, Daharde, Dandv, Daniel233, Daniel5127, DanielCD, Danny B-), Dar-Ape, Darekun, Darklich14, Darklilac, Dasyornis, Davewild, David D., David Wahler, DavidDidwin, DavidWBrooks, Davidiad, Daycd, Db099221, Dbenbenn, Dbrett480, Debresser, Decltype, Deflective, Delfigolo, Delta G, Demi s96, Demiurge, Deor, DerHexer, Derekrhu, Devil42400, Dfrg.msc, Dienamight, Dina, Dinsdalea, Dirty MF Harry, Discospinster, Dj Capricorn, Domminico, Doniago, Dori, DotComCairney, Download, Dr. Mandellez, DrMtl, Drdaveng, Dreadstar, Dreg743, Drphilharmonic, Dsinghs, Dullhunk, Dungodung, Dweller, Dwmyers, Długosz, EScribe, ESkog, Eaglestrike117, EarthPerson, EconoPhysicist, Ed, Edderso, Eddturtle, Eeekster, Eik Corell, Eleassar, Elmer Clark, EncycloPetey, Epbr123, Epipelagic, EqualMusic, Eric Kvaalen, EryZ, Etacar11, Ettrig, Etxrge, Eumeng.chong, Evercat, Everyking, Evil Monkey, Excirial, FF2010, Fahadsadah, Faithlessthewonderboy, FallingGravity, Feezo, Feniouk, Fenteany, Fermion, Fgbhrgthnrtgnb, FocalPoint, Fox6453, Fram, Freakofnurture, Fredrik, FreplySpang, Frymaster, Fullmetal123321, Funnybunny, Funper, Fvw, GHe, GTBacchus, Gacha1, Gadfium, Gaff, Gaius Cornelius, Gary King, GeeJo, GeeOh, Gekedo, Gene Nygaard, Germen, Gggh, Ghewgill, Giftlite, Gigemag76, Gilderien, Gilliam, Ginkgo100, Glenn, Goatman12, Govindjee, GregAsche, GregMinton, Gregturn, Grondemar, Gruzd, Guettarda, Guoguo12, Gwernol, HBNayr, HLHJ, Hadal, Haham hanuka, HalfShadow, Hans Dunkelberg, Hao2lian, Hardyplants, Hawaiian717, Hdt83, Hello32020, Henrik, Herbertxu, Heron, Hgrenbor, Hi112233, Hmains, Hometown, Hordaland, Hpfreak26, Hu12, Hunrizzo8, Hydrogen Iodide, II MusLiM HyBRiD II, ITOD, Iam angryrawr, Ian Pitchford, Iapetus, Icseaturtles, Ihope127, Imanoob69, Imi2012, Imogen89, InNuce, Incredio, Indon, Init, Intelligentsium, Interplanet Janet, Irishguy, Iron Dragon91, Ixfd64, J.delanoy, JDnCoke, JForget, JPaestpreornJeolhlna, JWSchmidt, JYolkowski, Jack No1, Jackgeddes1234, Jacob635, Jaknouse, JamesR777, Jamesscottbrown, Jamib0y, Jan1nad, Jasonphollinger, Jauhienij, Jaxl, Jay-Sebastos, Jazzy211, Jecar, Jeffq, Jennavecia, Jfg284, Jklin, Jlittlet, Jni, Joanjoc, JoaoRicardo, Johann Wolfgang, John, John of Reading, John254, Johna100, JonRichfield, Jonadin93, Jondel, JorisvS, Josh Grosse, Joshjoshjosh1, JoshuaZ, Jovianeye, Jpoelma13, Jrbouldin, Jrtayloriv, Jtdirl, Jugger90, Juhwade, Jusjih, Jwoodger, Jwray, K-UNIT, KFP, Kaabi, Kablammo, Kamal Chid, Kasper90, Kazvorpal, Kbh3rd, Kcordina, Keilana, Kevyn, Kim Bruning, King Toadsworth, King of Hearts, Kingpin13, Kopid03, Korg, Kungfuadam, Kungming2, Kupirijo, Kuru, Kww, Kylcon2, Kylemcinnes, L'Aquatique, LOL, La goutte de pluie, Lanoitarus, Larryorr, Laur2ro, Lawrenceuniversity1, Lemchesvej, Li-sung, Lightmouse, Lights, LikeLakers2, Likeabird, Lily15, Logan, Looxix, Lord 1284, Lord.lucan, Lotje, Luna Santin, Lupin, MC10, MER-C, MNAdam, MPF, Magister Mathematicae, Magnus Manske, Makemi, Malcolm, Malcolm Farmer, Malke 2010, Malkinann, Malljaja, Malo, Malomaboy06, Mani1, ManuelGR, Marj Tiefert, Martin451, Marysunshine, Master Jay, Mat-C, Materialscientist, Math Champion, Mattbr, Matthuxtable, MattieTK, Mav, Maxim, Mboverload, McSly, Mdavies 965, Me big fishy, Melchoir, Meltzerju, Mgiganteus1, Mhking, Michael C Price, Michael Hardy, Michael Ward, Mikael Häggström, Miketam, Mikimickz, Mimihitam, Minimac, Minna Sora no Shita, Mion, Mj455972007, Mlessard, Modster, Modulatum, Mogism, Molybdenumblue, Momo san, Mooinglemur, Morning277, Mrg024, Mrmonsterman, Mrrhum, Mrwarriorman, Munita Prasad, Munyanah, Mwarren us, Mygerardromance, Myrryam, NMChico24, NMFA Northern Eagle, NYMFan69-86, Nakon, Name b, NarSakSasLee, Narayanese, Natalie Erin, NativeForeigner, Nbauman, Nehrams2020, Neko-chan, NeoJustin, NewEnglandYankee, Neyne, Nibls, Nicholasink, Nick, Nick125, Nickresch, Nik42, NinaStarry, Nisadaninisa, Nivix, Nnewton, Nonagonal Spider, Nonluddite, Norrello, Nsk92, Nsteinberg, Nuno Tavares, Nuvitauy07, Nv8200p, O le gend, OMGHeHasAGun, Ocatecir, Oggie boggie in the loo, Ohnoitsjamie, Ohyeahbabe, Oiws, Okedem, Oliverc1991, Onco p53, Optichan, Ottawa4ever, Ouishoebean, OverlordQ, OwenX, Oxnora2012, Oxymoron83, P a k 2, P00100p, PDH, PaePae, Paine Ellsworth, Pajast, Pajz, Patrickdavidson, Paul August, Pavel Vozenilek, Pcruce, Peipei, Pekaje, Pengo, Pentawing, Perey, Persian Poet Gal, Peruvianllama, Petiatil, Pgk, Philip Trueman, Phoenix Hacker, Piano non troppo, Pierpontpaul2351, Pinethicket, Pinkadelica, Pinzo, Pipedreambomb, Pkoden, Pkpat2011, Platoface, Plociam, PoccilScript, PoeticX, Polyhedron, Pontifex101, Postdlf, Postglock, Primal400, Prince Godfather, Pro crast in a tor, Procrastinator, Prodego, Proguitarboy, Prunesqualer, Pvgetsome, Quadpus, Quiddity, Quintote, Qwerqwerqwer, Qxz, R Calvete, RJASE1, RJHall, RMFan1, Rachelpurdon, Racknack7824, Ram-Man, Rambatino, Ramcy, Randomnotice, Raven4x4x, Raymondwinn, Reach Out to the Truth, Reaper X, Red Herring, Red Winged Duck, RedWordSmith, RexNL, Rgamble, Rhysboy, Riader0, Rich Farmbrough, Rje, Rjwilmsi, Rmosler2100, Rmstock, Rob Hooft, Robchurch, Robert Thyder, Robsavoie, Rock2e, Ronhjones, RoyBoy, Rozzychan, Rrburke, Ryanrs, Ryantoh wiki, Ryulong, Saif11, Salsb, Samballance, Samsara, Samtux, Sango123, Sannse, Sarin123, Saseigel, Sbharris, Scarian, Sceptre, SchfiftyThree, Schlice, Scohoust, ScottSteiner, Sd card, Sdittami, Sdornan, Search4Lancer, SebastianHelm, Selket, Sengkang, Sergey WereWolf, Shadowjams, Shafei, Shanel, Shenme, Shiningstarr1997, Shorespirit, Shotwell, Silasmellor, Simpsons contributor, Sjakkalle, Skizzik, Skäpperöd, Slashme, SlimVirgin, Smartse, Smith609, Smokizzy, Snigbrook, Someguy1221, Sonett72, SonicAD, Soumyasch, SpaceFlight89, SparrowsWing, Spartan-James, SpeedyGonsales, Spiff, SpuriousQ, Srleffler, Staccatoque, Starzjake, Staticd, Stephenchou0722, Steveadcuk, Steven J. Anderson, Steven Zhang, Stevenmitchell, Stoic Squirrel, Storkk, Stroppolo, Studentne, SueHay, Sundar, Sunholm, Supernaturalist, Surtsicna, Sweet xx, SweetNeo85, Swmcd, Szajci, THEN WHO WAS PHONE?, Tackyleprechaun, Talrias, Tameeria, Tarquin, Tastycheeze, Tattoo275, Tawker, Tbhotch, Tchaika, Tdslk, TedE, Teh wise one, Tellyaddict, Tero, TestPilot, TexasAndroid, Thanatos666, Thatotherdude, The Catfish, The Earwig, The Thing That Should Not Be, The sock that should not be, The wub, TheArmadillo, TheTito, Thingg, Tide rolls, TigerShark, Tikoo, Tim Q. Wells, TimVickers, Tito4000, Tn4196, Toadams, Tobby72, Tom Pippens,

Tomaschwutz, Tommy2010, Torahjerus14, Touchstone42, Towerofthunder, Tpbradbury, Tra, Tradnor, Trappist, Traroth, Treisijs, Tristanb, Troy 07, Twirligig, U.S.Vevek, Uiteoi, Ulric1313, Uncle Dick, Unclepea, Unyoyega, Upeksha112, Usdi, User A1, Username314, Valérie75, Van helsing, Vanished user, Vary, Vices, Viper777, VolcomRUXC, Vortexrealm, Vsmith, Vuvar1, Waggers, WatermelonPotion, Watplay, Wavelength, Wejer, WelshMatt, White.matthew.09, Whosyourjudas, Wiitennisxd, WikHead, Wiki alf, WikiLaurent, Wikidenizen, Wikidudeman, WikipedianMarlith, William Avery, Wimt, Wine Guy, Wk muriithi, Woohookitty, Woudloper, WriterHound, Wutsje, Xchbla423, Xezbeth, Xiong Chiamiov, Xoddam, Yamaguchi, Yamamoto Ichiro, Yerpo, Yikrazuul, Yon Yonson, Z 153, Zack, Zap Rowsdower, Zarniwoot, Zedla, Zepheus, Zfr, ZooFari, Zorahia, Zvn, Zxoxm, , 2161 anonymous edits

Glycolysis *Source*: http://en.wikipedia.org/w/index.php?title=Glycolysis *Contributors*: -), 2001:388:608C:4950:7D05:EF72:29BB:C100, Abarry, Addihockey10, Adrian J. Hunter, Agpetern, Aka042, Albacore, Alex F., AlexCruise, Alhead, Alice.odin, Allstarecho, Alnokta, AlphaEta, Alteripse, Amdurbin, Amkilpatrick, Andre Engels, Andres, Andrewpmk, Anetode, Arcadian, ArglebargleIV, Arnavchaudhary, Arthena, Asamos47, AubreyEllenShomo, AxelBoldt, Barticus88, Bensaccount, Bento00, Bogey97, Bongwarrior, BorisTM, Boytae, Brenont, BrightStarSky, Bryan Derksen, Bucketsofg, Buttholationszee, CBSB, CTZMSC3, Calvin 1998, Can't sleep, clown will eat me, CanisRufus, Cburnett, Ceyockey, Cheez277, Choij, Chris the speller, ChrisGualtieri, Clicketyclack, Colin, CommonsDelinker, Conversion script, Courserepjoe, Ctbolt, Cuculus, Cyberman, D-rew, Darsie, David D., Daycd, Dburson, Delldot, Delta G, Demian12358, Dempsey1717, DevilsAdvocate, Diberri, Dlamming, Dmdukes, Dnvrfantj, Dono, Drphilharmonic, Dude1818, Dullhunk, EconomicsGuy, Ekton, Element16, Elonka, Elronxenu, Enviroboy, Epbr123, Erianna, EricWesBrown, Etxrge, Evergreen vsh007, Exor674, Fenteany, FreplySpang, Friginator, Fuzlyssa, Fvasconcellos, G00nsf, G3pro, Gaius Cornelius, Gcm, Geoking66, Giftlite, Gilliam, Gimboid13, Glane23, Graham87, GraybeardBiochemist, Gtstricky, Gurch, Gurubrahma, Hadal, Harbinary, HexaChord, HoodedMan, Ibrmrn3000, Igotlotsaquarters, Ilphin, Isnow, Iwilcox, J G Campbell, J. Finkelstein, J.delanoy, JDoorjam, JForget, JHunterJ, JTN, JWSchmidt, JaGa, Jack the Stripper, Jack.schonbrun, Jag123, Jakob Theorell, Jauhienij, Jeff G., Jeffscott007, JeramieHicks, Jeremy.d.morgan, Jimhsu77479, Jmarchn, Jmauser, Joehall45, JohnnyCalifornia, JohnyAbb, Jojit fb, Jonas094, JonathonSimister, Josh Cherry, Jotomicron, Jrockley, JuanitaJP, Jusses2, K. Hiippari, Kaarel, Kandar, KathrynLybarger, Khullah, Kirez, Kjaergaard, Kramtark, KristinaHanspers, Krylonblue83, LHcheM, Laportechicago, Larry_Sanger, Laurenferruccio, Lentower, Lexivore, Lightmouse, Lilac Soul, Logan, Lugia2453, Magioladitis, Magister Mathematicae, Magnus Manske, Malcolm Farmer, Manticore, Marc Isaacs, Marcelo1229, Marj Tiefert, Markacohen, Maximaximax, Mean as custard, Meco, Meggar, Michall, MicroProf, Microbrain, Mifter, Miguel.mateo, Mikael Häggström, Mikechul, Mindmatrix, Miquonranger03, Mohawkjohn, Mpatel, Mplona, Munkee, N5iln, Narayanese, Navdar, Naveen57, Niccus, NightFalcon90909, Northumbrian, Nuttcrakerr, Oatmeal batman, OccamzRazor, OliAtlason, Opabinia regalis, Pakaran, Paulwithap, Pedvi, Pelirojopajaro, PhilANThropiSt, Pinethicket, PloniAlmoni, Plrk, Pmokeefe, Polypipe Wrangler, Popnose, Prav001, Pred7799, RainbowOfLight, Rainbowwrasse, Ranze, Ravnoor, Recnilgiarc, Reedy, Registrar, RetiredWikipedian789, Rhodekyll, Richard Arthur Norton (1958-), Richard001, Righthing, Rjwilmsi, Roland Kaufmann, Ronhjones, Rozzychan, Ruzha, RyanGerbil10, Ryanreednro, SCEhardt, Saravask, Sasuke Sarutobi, Satellizer, SchwarzeMelancholie, Scohoust, Screensaver7, Scwlong, Selket, Sfan00 IMG, Sfxdude, Shalom Yechiel, Shoeofdeath, Sitush, Slashme, Smooth O, Soakologist, Soliloquial, SouthernNights, Spamburgler, Spencerw, Speshuldusty, Stepa, Super tom2, Superdoggy, Svetlana Miljkovic, Swmmr1928, Szxd, Tabletop, Tekks, The Rambling Man, The Thing That Should Not Be, Thehelpfulone, TicketMan, TimVickers, Tito4000, Tonyest, Trnoriega, TsunadeSama, TutterMouse, Uhai, Una Smith, Use the force, Vin268, Widr, Wiki alf, Wikieditor06, Williamb, Wimt, Wknight94, Wolfinbm, Yappy2bhere, YassineMrabet, Yerpo, Yraser, Yuorme, Z0OMD, Zargulon, Zephyris, Zoicon5, 580 anonymous edits

Oxidative_phosphorylation *Source*: http://en.wikipedia.org/w/index.php?title=Oxidative_phosphorylation *Contributors*: 2Plus2Is4, Abstraktn, Adenosine, Alan Liefting, Alnokta, Antelan, Anton Gutsunaev, Arcadian, Art LaPella, AshLin, BD2412, Baharl01, Bensaccount, Bobo192, Borgx, BorisTM, Brenont, Brighterorange, Bryan Derksen, Can't sleep, clown will eat me, Chris Rodgers, Clicketyclack, CommonsDelinker, Conversion script, CopperKettle, Crazycomputers, Cst17, DJ Clayworth, DMacks, Danny, Davaith, David D., Dcirovic, Dj Capricorn, Drphilharmonic, Dwmyers, EerieNight, El C, Epbr123, Eribro, Espresso Addict, Etxrge, Eubulides, Faramir333, Fvasconcellos, Gaius Cornelius, Graham87, GraybeardBiochemist, Habj, Harryboyles, Hodja Nasreddin, IKenny, IRP, Ijustam, Imjustmatthew, J.delanoy, JWSchmidt, Jag123, Jamouse, Jesse V., Josh Cherry, Jpg, Julianonions, Jwollbold, K Y Nelson, Kakli, Kozuch, Laurip, LedgendGamer, Leonard^Bloom, Leptictidium, Lesviolonsdautomne, LightBringer, Lilac Soul, MITBeaverRocks, Macowell, Margaretabuttery, Marj Tiefert, Materialscientist, Maxim Gavrilyuk, Michael Devore, Mikael Häggström, Mwucherer, Narayanese, Nhandler, Nono64, Nwbeeson, ParticleMan, Persian Poet Gal, Philgoetz, Piledhigheranddeeper, Pinkadelica, Plugs0, RDBrown, Rakwiki, Rich Farmbrough, Richard001, Richard0612, Rjwilmsi, Semitallsmalls, Shahriyar alavi, Simpsons contributor, Slowpokeiv, Smokefoot, Smurrayinchester, Some jerk on the Internet, Some standardized rigour, StaticGull, Stepa, Stewartadcock, StradivariusTV, Suruena, Tameeria, Tedtoal, TimVickers, Treisijs, Ttownfeen, Ulisse0, Verticalstall, Wavelength, Wmpearl, WolfmanSF, WriterHound, Xp54321, Zephyris, שמוליק, 166 anonymous edits

Circadian_clock *Source*: http://en.wikipedia.org/w/index.php?title=Circadian_clock *Contributors*: Boghog, Cng168, Daleslimnsaw, DavidWBrooks, Estel, Fvw, Gorton k, HansPopper, Hordaland, Mon3oturf, Rbrwr, TracyRenee, Woohookitty, 7 anonymous edits

Nitrogen_assimilation *Source*: http://en.wikipedia.org/w/index.php?title=Nitrogen_assimilation *Contributors*: Cacycle, DanielCD, Illia Connell, Klemen Kocjancic, Lexor, LilHelpa, Magioladitis, Marcus2, Neodop, Onco p53, Rkitko, Sidelight12, Staticd, Strait, Wavelength, 15 anonymous edits

Stress_(biology) *Source*: http://en.wikipedia.org/w/index.php?title=Stress_%28biology%29 *Contributors*: 1000Faces, 1945AlphaTeam, 28421u2232nfenfcenc, 4twenty42o, 5 albert square, A. B., A.Z., Aadityabhatia, Aaron Kauppi, Aaron Schulz, Abductive, Abeg92, Adam mugliston, Adwiii, Agent gruer, Aitias, Akhil1904, Alansohn, Alcor2010, Ale jrb, Alex.tan, AlexOlsen, Alexoprea, Allwynmascgmail, Allycat, Altenmann, Amatulic, Amitkochhar, Amorveritas, AnatomyProf0001, Andrewpmk, Andycjp, Angr, Animeronin, Animum, Anlace, Anna Lincoln, Anthonyhcole, Apoptic, Applebee421, Ariamaki, Ariseminor, Arthena, Aschu, Atcack, Atif.t2, Atroche, Auntof6, Avialae, Avigdor6, AxelBoldt, Barrylb, Bearian, Beebeefails, Beland, Belovedfreak, Ben Tibbetts, Bento00, Betterusername, Between My Ken, Bhny, Bige1977, Bigfastmercury, Biker Biker, Bility, Binary TSO, Bkabaker, Bkell, Blanchardb, Bobblehead, Bobo192, Bodnotbod, Bogglevit, Bonadea, Bonedoc007, BoundaryRider, Brad7777, Bradeos Graphon, Brenont, Brian Crawford, BrianMSweis, Brianbudgell, Brianga, Briern, BrokenSegue, Buchanan-Hermit, Bwefler, CALR, CZ2, Cablehorn, Caltas, CardinalDan, Carsgobroombroom, Casey Abell, Caue.cm.rego, Cdang, Cecity, Cedders, Cflex0010, Chasingsol, Cheifsguy, Chloem86, Chochopk, Chowbok, Chris 73, ChrisKees5, Ckatz, Clam0p, Clayoquot, Cleared as filed, Closedmouth, Cma53, Cminard, Copywriterpro, Cory Kohn, Courcelles, Crimsoncard641, CrookedAsterisk, CrownprinceJohannes, Cruciatum, CryptoDerk, Csbrewfisher, Cyberpower678, DabMachine, Daltonj06, DanielCD, Daonguyen95, Darkemagik, Darktrail345, Darth Panda, Dastrophy, David McQuillan, David spector, Deabruzo, DeadEyeArrow, Debra1952, Debresser, Deli nk, Deodar, Der Golem, DerHexer, Dewy94, Dgw, Dharris, Dimitri108, DoubleBlue, Dougmcdonell, Download, DrKiernan, Dreadstar, Dreamyshade, Drphilharmonic, Drtimlow, Dubious Irony, Dust Filter, Dustmagic, Dysepsion, E.Mitsui, EJF, Ego White Tray, Egsan Bacon, Eleassar777, Eleschinski2000, Elf, Enviroboy, Epbr123, Espoo, Etrielle, EunGi, EverSince, Excirial, Exnerblc, Eyesnore, Eyu100, FT2, Face, Falcon8765, Felicity86, Figmented, Fireice, Firetrap9254, Firien, Fishermansbay, Flameninja311, Flopster2, Fraggle81, Fram, Freedomtalks, FreplySpang, Funandtrvl, Gaggafuto, Geke, Glane23, Gleng, Glenn, Goeagles4321, Gogo Dodo, GraemeL, Graham87, Grantandersonuk, Gridge, Gscshoyru, Gtstricky, Gurch, Hadal, Hamiltondaniel, Hammersoft, Hanahime, HappyInGeneral, Harriv, Hathserv, Head on a silver plate, Helikophis, Hello cello, Hepuli, Heroeswithmetaphors, Heron, Hhgfhfh, Hibana, Hobawido, Hodja Nasreddin, Hu, Hucz, Hurdles87, Hydroflexology, I dream of horses, Iampoodlebunny, IddoGenuth, Ideal gas equation, Immunize, Insanity Incarnate, Invader05, Invertzoo, Iridescent, Isopropyl, Iss246, Ivan Kirillov, Ivaniceconc, J. Straub, J.delanoy, JForget, JHunterJ, JMS Old Al, JPBarbuto, JRHorse, JYolkowski, JaGa, Jackson Peebles, Jakezing, James086, Jamiek82, Jandalhandler, Japanese Searobin, Jarble, Jauhienij, Jaxl, Jaymoon12, Jcbutler, Jetekus, Jfdwolff, John Barbuto, MD, JohnCD, Johnuniq, Johnyoosuk, JonHarder, Jonathan321, JorisvS, Joyson Prabhu, Jprg1966, Jschnur, Jtneill, Juliancolton, Junglejill, KGasso, Kalaiarasy, Karch, Karen Johnson, Karlhahn, Kars777, Kesh, Khazar2, Kigoe, Kingpin13, Koavf, Kubigula, Kubra, Kurieeto, Kyoko, L Kensington, LAX, LOL, La goutte de pluie, Lacort, Larry laptop, Leibniz, LeilaniLad, Leništudent, Leothelion302000, Lgh, Lightmouse, LilHelpa, Linnell, Liquidcentre, LittleOldMe, Loadmaster, Looie496, Lothar von Richthofen, Lova Falk, Love ShayBishop, Lugia2453, M0rphzone, MER-C, Mad Greg, Mahonk19, Mani1, Manuel Anastácio, MarkBuckles, Martinq22, Materialscientist, Mathieu2003, Matthewrlee, Mattisse, Maurice Carbonaro, Maxí, Meegs, Mentifisto, Mentisock, Michael Henchard, Michael2, MichaelExe, Mig77, Mikael Häggström, MikeVitale, MindstormsKid, Mindygifford, Mmminwoo99, Mmxx, Molkhal, Moshe Constantine Hassan Al-Silverburg, Mr-Natural-Health, Mr.Z-man, MrOllie, Mrb03, Mrmusashi, Mushroom, Mushroom Pi, Muéro, N8da4m, Nastajus, Nathan.tay, Natural Cut, Naviguessor, NawlinWiki, Nbarth, Neezes, NeilUK, Neko-chan, Neocobalt, Nick C, Nick Number, Nigelloring, Nightngle, Nigosh, Nihiltres, Nlamore, Noah Salzman, Nonnb, NotAnonymous0, NotWith, Nsamuels, Ntse, Nutriveg, Ohnoitsjamie, Omnechist, OnBeyondZebrax, OnaTutors, Onceonthisisland, Orphan Wiki, Ortolan88, Outlook, OverlordQ, Owen, Oxymoron83, PGWG, Pajast, Pariah, Pathoschild, Patrick, Paul A, Pavel Vozenilek, Pcb21, Pedant17, Peter-ZA, Pgan002, Pgreenfinch, Phantomsnake, Phantomsteve, Philip Trueman, Pinethicket, Pmcchesn, Positivenegative, Prabanan, Punkymonkey987, Pythonesque, Qwerty Binary, Qwyrxian, R Lowry, R'n'B, RWardy, RandomP, Randomwellwisher, Ratarsed, Rbbloom, Rcarlosagis, RedBeach7, Reinis, RenamedUser01302013, RetiredUser124642196, RexNL, RichardF, Ricky81682, Rider tiger, Rigadoun, Rixs, Rjwilmsi, Robert Daoust, Rocketrod1960, Rodii, Rosser1954, Roux-HG, RoyBoy, Rrburke, RuED, Russell E, SWAdair, Saint41, Saminshah, SandyGeorgia, Sannse, Sav vas, Sciencewatcher, Sepper, Septegram, Serenasenses, Shadowjams, Shandithrt, Shiaru, Shnuis, Shriniwaskashalikar, Silly rabbit, SimenH, SirNuke, Skittleys, Skizzik, Skumarla, Skyfire808, Smalltowngirl, SoCal, SoCalSuperEagle, Soot21000003333, Sophus Bie, South Bay, Spatranobis, Spazure, Squids and Chips, StephenBuxton, SteveNash11, Stevenfruitsmaak, Steventanoto, StormyJerry, Strike Eagle, Stronglifts, Stryn, Sutherford, THE KING, Tabletop, Tanvir Ahmmed, Tapdancer2000, Taroaldo, TarseeRota, Tedeh, Tekaphor, TexasAndroid, Texture, Tgeairn, The Anome, The Man in Question, The Rationalist, The Thing That Should Not Be, The Transhumanist, TheAMmollusc, Thecheesykid, Thedreamdied, ThedukeSL, Theguyinblue, Thenewguy34, Thingg, Thorn breaker, Thseamon, Tide rolls, Tim Starling, Timflarp220, TinyE, Tiptoety, Titoxd, Tobias Bergemann, Tolly4bolly, Tommy2010, Trevon watford, Tripod2282, Troust, Truthbody, Tryptofish, Twas Now, Uthbrian, Uusijani, Versageek, Vexorg, Vinsum21, Vrenator, WLU, Wavelength, Wblakesx, Weregerbil, WhatamIdoing, Wholelife, WikiOriginal, Wikilibrarian, Wikipelli, Willdude 132, Wingspeed, Winterst, Wintran, Wk muriithi, Wmahan, Wouterstomp, WpZurp, WriterHound, Wyckyd Sceptre, XMog, Xionbox, Yahya Abdal-Aziz, Yamamoto Ichiro, Yoenit, Yogainfo, Zoharadner, Zsinj, , 1145 anonymous edits

Plastid *Source*: http://en.wikipedia.org/w/index.php?title=Plastid *Contributors*: A876, AdamRetchless, Ahoerstemeier, Animum, Anonymous Dissident, Arbitrarily0, Atubeileh, Audriusa, AxelBoldt, Banus, Bejnar, Beland, Blue Danube, Bomac, Calvin 1998, ClockworkSoul, Closedmouth, CommonsDelinker, DabMachine, Dave3457, Davepape, Dawnseeker2000, Daycd, Digitalgadget, Dixi, Drphilharmonic, Dullhunk, Elaragirl, EncycloPetey, Erebus Morgaine, Excirial, Extremophile, Fetofs, Filnik, Freakofnurture, G.Voelcker, GeTAboozE, Giftlite, Glenn, Gogo Dodo, Gracenotes, Graminophile, Hdt83, Hindarion, Ipodamos, Iztwoz, JamesAM, Jlundell, Johnnybaseball142, Jonadin93, Josh Grosse, Jossi, Jredmond, Jwrosenzweig, Kelvinsong, KeybladeSephi, Kingpin13, Kubigula, LadyofHats, Larwe, Lavateraguy, LeaveSleaves, Lilious, Luis Fernández García, Lupin, Luuva, Magnus Manske, Manop, Materialscientist, Mcherrill, Meddoc13, Mgiganteus1, Munita Prasad, NickCT, Nilmerg, OakRunner, OrenBochman, Osborne, PAvdK, PaddyM, Paine Ellsworth, Palica, PeepP, Philip Trueman, Pinethicket, Plantsurfer, Proofreader77, RDBrown, RHaden, Razorflame, Riley Huntley, Ronald.snijder, Sakkura, SaveThePoint, Sevrin nih, Smartse, Snigbrook, Spamburgler, SpuriousQ, Steven Zhang, Sylverfysh, Tagishsimon, Tcncv, Tentinator, Thadius856, The Thing That Should Not Be, TheAlphaWolf, Troy 07, TutterMouse, Vojtech.dostal, Wasell, Wayne Slam, WereSpielChequers, Whoop whoop pull up, Wikier.ko, WillowW, WriterHound, Xavier Thibault, Xx 0wn3ed xx, Yubcvan, Zfr, Zorahia, Zzuuzz, 243 anonymous edits

Image Sources, Licenses and Contributors

file:Tolypothrix (Cyanobacteria).JPG *Source*: http://en.wikipedia.org/w/index.php?title=File:Tolypothrix_(Cyanobacteria).JPG *License*: unknown *Contributors*: User:Matthewjparker

File:2010 Filamentous Cyanobacteria Bloom near Fiji.jpg *Source*: http://en.wikipedia.org/w/index.php?title=File:2010_Filamentous_Cyanobacteria_Bloom_near_Fiji.jpg *License*: unknown *Contributors*: Norman Kuring

File:CyanobacteriaColl1.jpg *Source*: http://en.wikipedia.org/w/index.php?title=File:CyanobacteriaColl1.jpg *License*: unknown *Contributors*: User:Fice

File:Clyndrospermum (Cyanobacteria).JPG *Source*: http://en.wikipedia.org/w/index.php?title=File:Clyndrospermum_(Cyanobacteria).JPG *License*: unknown *Contributors*: User:Matthewjparker

File:Cyanobacterium-inline.svg *Source*: http://en.wikipedia.org/w/index.php?title=File:Cyanobacterium-inline.svg *License*: unknown *Contributors*: User:Kelvinsong

File:ZebRivStromatolites 2010.1.jpg *Source*: http://en.wikipedia.org/w/index.php?title=File:ZebRivStromatolites_2010.1.jpg *License*: unknown *Contributors*: User:NimbusWeb

File:OncolitesAlamoBreccia.jpg *Source*: http://en.wikipedia.org/w/index.php?title=File:OncolitesAlamoBreccia.jpg *License*: unknown *Contributors*: User:Wilson44691

File:Haeckel arbol bn.png *Source*: http://en.wikipedia.org/w/index.php?title=File:Haeckel_arbol_bn.png *License*: unknown *Contributors*: User:Luis_Fernández_García

File:Blue-green algae cultured in specific media.jpg *Source*: http://en.wikipedia.org/w/index.php?title=File:Blue-green_algae_cultured_in_specific_media.jpg *License*: unknown *Contributors*: User:JDP90

File:Spirulina tablets.jpg *Source*: http://en.wikipedia.org/w/index.php?title=File:Spirulina_tablets.jpg *License*: unknown *Contributors*: Original uploader was Perdita at the English Wikipedia

File:CC-BY-icon-80x15.png *Source*: http://en.wikipedia.org/w/index.php?title=File:CC-BY-icon-80x15.png *License*: unknown *Contributors*: Creative Commons

File:Auto-and heterotrophs.png *Source*: http://en.wikipedia.org/w/index.php?title=File:Auto-and_heterotrophs.png *License*: unknown *Contributors*: User:Mikael Häggström

File:Troph flowchart.png *Source*: http://en.wikipedia.org/w/index.php?title=File:Troph_flowchart.png *License*: unknown *Contributors*: User:Wickey-nl

File:Photosynthesis.gif *Source*: http://en.wikipedia.org/w/index.php?title=File:Photosynthesis.gif *License*: unknown *Contributors*: User:At09kg

File:Photosynthesis equation.svg *Source*: http://en.wikipedia.org/w/index.php?title=File:Photosynthesis_equation.svg *License*: unknown *Contributors*: User:Zoofari

File:Seawifs global biosphere.jpg *Source*: http://en.wikipedia.org/w/index.php?title=File:Seawifs_global_biosphere.jpg *License*: unknown *Contributors*: Julia W, Luis Fernández García, Tano4595, TheDJ, Túrelio, Yikrazuul, 1 anonymous edits

File:Simple photosynthesis overview.svg *Source*: http://en.wikipedia.org/w/index.php?title=File:Simple_photosynthesis_overview.svg *License*: unknown *Contributors*: User:Maveric149, User:Yerpo

File:Chloroplast.svg *Source*: http://en.wikipedia.org/w/index.php?title=File:Chloroplast.svg *License*: unknown *Contributors*: User:Emmanuel.boutet

File:Thylakoid membrane.png *Source*: http://en.wikipedia.org/w/index.php?title=File:Thylakoid_membrane.png *License*: unknown *Contributors*: Original uploader was Tameeria at en.wikipedia

File:Z-scheme.png *Source*: http://en.wikipedia.org/w/index.php?title=File:Z-scheme.png *License*: unknown *Contributors*: Enochlau, Mike.lifeguard, The cat, 1 anonymous edits

File:Calvin-cycle4.svg *Source*: http://en.wikipedia.org/w/index.php?title=File:Calvin-cycle4.svg *License*: unknown *Contributors*: User:Adenosine

File:HatchSlackpathway2.svg *Source*: http://en.wikipedia.org/w/index.php?title=File:HatchSlackpathway2.svg *License*: unknown *Contributors*: User:Adenosine, User:Jamouse

File:Measuring photosynthetic efficiency using an LCpro-SD.jpg *Source*: http://en.wikipedia.org/w/index.php?title=File:Measuring_photosynthetic_efficiency_using_an_LCpro-SD.jpg *License*: unknown *Contributors*: User:Steveadcuk

File:Plagiomnium affine laminazellen.jpeg *Source*: http://en.wikipedia.org/w/index.php?title=File:Plagiomnium_affine_laminazellen.jpeg *License*: unknown *Contributors*: User:Fabelfroh

File:Melvin Calvin.jpg *Source*: http://en.wikipedia.org/w/index.php?title=File:Melvin_Calvin.jpg *License*: unknown *Contributors*: "Photolab"

File:Leaf 1 web.jpg *Source*: http://en.wikipedia.org/w/index.php?title=File:Leaf_1_web.jpg *License*: unknown *Contributors*: Jon Sullivan

File:Photorespiration.svg *Source*: http://en.wikipedia.org/w/index.php?title=File:Photorespiration.svg *License*: unknown *Contributors*: User:Yikrazuul

File:Glycolysis2.svg *Source*: http://en.wikipedia.org/w/index.php?title=File:Glycolysis2.svg *License*: unknown *Contributors*: User:JohnyAbb

File:D-glucose wpmp.png *Source*: http://en.wikipedia.org/w/index.php?title=File:D-glucose_wpmp.png *License*: unknown *Contributors*: Richard Wheeler (Zephyris)

File:Biochem reaction arrow forward NNNN horiz med.png *Source*: http://en.wikipedia.org/w/index.php?title=File:Biochem_reaction_arrow_forward_NNNN_horiz_med.png *License*: unknown *Contributors*: Richard Wheeler (Zephyris)

File:Pyruvate2 wpmp.png *Source*: http://en.wikipedia.org/w/index.php?title=File:Pyruvate2_wpmp.png *License*: unknown *Contributors*: User:Iwilcox, User:NEUROtiker

File:Glycolysis.jpg *Source*: http://en.wikipedia.org/w/index.php?title=File:Glycolysis.jpg *License*: unknown *Contributors*: Tekks (talk). Original uploader was Tekks at en.wikipedia

image:D-glucose wpmp.png *Source*: http://en.wikipedia.org/w/index.php?title=File:D-glucose_wpmp.png *License*: unknown *Contributors*: Richard Wheeler (Zephyris)

image:Alpha-D-glucose-6-phosphate wpmp.png *Source*: http://en.wikipedia.org/w/index.php?title=File:Alpha-D-glucose-6-phosphate_wpmp.png *License*: unknown *Contributors*: Richard Wheeler (Zephyris)

image:Biochem_reaction_arrow_forward_YYNN_horiz_med.png *Source*: http://en.wikipedia.org/w/index.php?title=File:Biochem_reaction_arrow_forward_YYNN_horiz_med.png *License*: unknown *Contributors*: Richard Wheeler (Zephyris)

image:Beta-D-fructose-6-phosphate wpmp.png *Source*: http://en.wikipedia.org/w/index.php?title=File:Beta-D-fructose-6-phosphate_wpmp.png *License*: unknown *Contributors*: Richard Wheeler (Zephyris)

image:Biochem_reaction_arrow_reversible_NNNN_horiz_med.png *Source*: http://en.wikipedia.org/w/index.php?title=File:Biochem_reaction_arrow_reversible_NNNN_horiz_med.png *License*: unknown *Contributors*: Richard Wheeler (Zephyris)

image:beta-D-fructose-1,6-bisphosphate_wpmp.png *Source*: http://en.wikipedia.org/w/index.php?title=File:Beta-D-fructose-1,6-bisphosphate_wpmp.png *License*: unknown *Contributors*: Richard Wheeler (Zephyris)

image:D-glyceraldehyde-3-phosphate wpmp.png *Source*: http://en.wikipedia.org/w/index.php?title=File:D-glyceraldehyde-3-phosphate_wpmp.png *License*: unknown *Contributors*: Richard Wheeler (Zephyris)

image:glycerone-phosphate_wpmp.png *Source*: http://en.wikipedia.org/w/index.php?title=File:Glycerone-phosphate_wpmp.png *License*: unknown *Contributors*: Richard Wheeler (Zephyris)

image:1,3-bisphospho-D-glycerate.svg *Source*: http://en.wikipedia.org/w/index.php?title=File:1,3-bisphospho-D-glycerate.svg *License*: unknown *Contributors*: User:Panoramix303

image:Biochem_reaction_arrow_reversible_YYYY_horiz_med.png *Source*: http://en.wikipedia.org/w/index.php?title=File:Biochem_reaction_arrow_reversible_YYYY_horiz_med.png *License*: unknown *Contributors*: Richard Wheeler (Zephyris)

image:3-phospho-D-glycerate_trulyglycerate_wpmp.png *Source*: http://en.wikipedia.org/w/index.php?title=File:3-phospho-D-glycerate_trulyglycerate_wpmp.png *License*: unknown *Contributors*: User:Iwilcox

image:2-phospho-D-glycerate_wpmp.png *Source*: http://en.wikipedia.org/w/index.php?title=File:2-phospho-D-glycerate_wpmp.png *License*: unknown *Contributors*: Richard Wheeler (Zephyris)

image:phosphoenolpyruvate_wpmp.png *Source*: http://en.wikipedia.org/w/index.php?title=File:Phosphoenolpyruvate_wpmp.png *License*: unknown *Contributors*: Richard Wheeler (Zephyris)

image:Biochem_reaction_arrow_reversible_NYYN_horiz_med.png *Source*: http://en.wikipedia.org/w/index.php?title=File:Biochem_reaction_arrow_reversible_NYYN_horiz_med.png *License*: unknown *Contributors*: Richard Wheeler (Zephyris)

image:pyruvate_wpmp.png *Source*: http://en.wikipedia.org/w/index.php?title=File:Pyruvate_wpmp.png *License*: unknown *Contributors*: Richard Wheeler (Zephyris)

File:Glycolysis free energy changes.svg *Source*: http://en.wikipedia.org/w/index.php?title=File:Glycolysis_free_energy_changes.svg *License*: unknown *Contributors*: User:Popnose

File:Hexokinase B 1IG8 wpmp.png *Source*: http://en.wikipedia.org/w/index.php?title=File:Hexokinase_B_1IG8_wpmp.png *License*: unknown *Contributors*: Original uploader was Zephyris at en.wikipedia

File:Phosphofructokinase 6PFK wpmp.png *Source*: http://en.wikipedia.org/w/index.php?title=File:Phosphofructokinase_6PFK_wpmp.png *License*: unknown *Contributors*: Original uploader was Zephyris at en.wikipedia

File:Pyruvate Kinase 1A3W wpmp.png *Source*: http://en.wikipedia.org/w/index.php?title=File:Pyruvate_Kinase_1A3W_wpmp.png *License*: unknown *Contributors*: Original uploader was Zephyris at en.wikipedia

Image:WP534.png *Source*: http://en.wikipedia.org/w/index.php?title=File:WP534.png *License*: unknown *Contributors*: KristinaHanspers

File:Magnify-clip.png *Source*: http://en.wikipedia.org/w/index.php?title=File:Magnify-clip.png *License*: unknown *Contributors*: User:Erasoft24

File:Mitochondrial electron transport chain—Etc4.svg *Source*: http://en.wikipedia.org/w/index.php?title=File:Mitochondrial_electron_transport_chain—Etc4.svg *License*: unknown *Contributors*: User:Fvasconcellos

File:Ubiquinone–ubiquinol conversion.svg *Source*: http://en.wikipedia.org/w/index.php?title=File:Ubiquinone–ubiquinol_conversion.svg *License*: unknown *Contributors*: User:Fvasconcellos

File:Complex I.svg *Source*: http://en.wikipedia.org/w/index.php?title=File:Complex_I.svg *License*: unknown *Contributors*: User:Fvasconcellos

File:Complex II.svg *Source*: http://en.wikipedia.org/w/index.php?title=File:Complex_II.svg *License*: unknown *Contributors*: User:Fvasconcellos

File:Complex III reaction.svg *Source*: http://en.wikipedia.org/w/index.php?title=File:Complex_III_reaction.svg *License*: unknown *Contributors*: User:Fvasconcellos

File:Complex IV.svg *Source*: http://en.wikipedia.org/w/index.php?title=File:Complex_IV.svg *License*: unknown *Contributors*: User:Fvasconcellos

File:ATPsyn.gif *Source*: http://en.wikipedia.org/w/index.php?title=File:ATPsyn.gif *License*: unknown *Contributors*: Original uploader was TimVickers at en.wikipedia

File:Nitrate ion balance-variants.svg *Source*: http://en.wikipedia.org/w/index.php?title=File:Nitrate_ion_balance-variants.svg *License*: unknown *Contributors*: User:Staticd

File:Rotating brain colored.gif *Source*: http://en.wikipedia.org/w/index.php?title=File:Rotating_brain_colored.gif *License*: unknown *Contributors*: lifesciencedb. Original uploader was BrianMSweis at en.wikipedia

File:Gray 111 - Vertebral column-coloured.png *Source*: http://en.wikipedia.org/w/index.php?title=File:Gray_111_-_Vertebral_column-coloured.png *License*: unknown *Contributors*: User: Uwe Gille

File:Illu adrenal gland.jpg *Source*: http://en.wikipedia.org/w/index.php?title=File:Illu_adrenal_gland.jpg *License*: unknown *Contributors*: Arcadian, AxelBoldt, Jmarchn, Photohound

File:Cortisol2.svg *Source*: http://en.wikipedia.org/w/index.php?title=File:Cortisol2.svg *License*: unknown *Contributors*: Calvero.

File:HPA Axis Diagram (Brian M Sweis 2012).png *Source*: http://en.wikipedia.org/w/index.php?title=File:HPA_Axis_Diagram_(Brian_M_Sweis_2012).png *License*: unknown *Contributors*: User:BrianMSweis

Image:General Adaptation Syndrome.jpg *Source*: http://en.wikipedia.org/w/index.php?title=File:General_Adaptation_Syndrome.jpg *License*: unknown *Contributors*: User:David McQuillan

File:PTSD stress brain.gif *Source*: http://en.wikipedia.org/w/index.php?title=File:PTSD_stress_brain.gif *License*: unknown *Contributors*: National Institutes of Health

Image:Plastids types en.svg *Source*: http://en.wikipedia.org/w/index.php?title=File:Plastids_types_en.svg *License*: unknown *Contributors*: User:LadyofHats

MIX
Papier aus verantwortungsvollen Quellen
Paper from responsible sources
FSC® C105338

Printed by Books on Demand GmbH, Norderstedt / Germany